W9-AEO-140

Technology Forces at Work

Profiles of Environmental Research and Development at DuPont, Intel, Monsanto, and Xerox

Susan Resetar

with

Beth E. Lachman

Robert Lempert

Monica M. Pinto

Supported by the

Office of Science and Technology Policy

RAND

SCIENCE AND TECHNOLOGY POLICY INSTITUTE

The environmental literature recognizes the importance of involving multiple stakeholders in the environmental policy development process. Stakeholders include a diverse set of individuals and organizations—local citizens and community groups, consumers, environmental groups, industry, individual companies, shareholders, all levels of government and tribes, etc. In turn, each of these will have various perspectives on environmental risk, priorities, costs and benefits, etc.

One area of environmental policy that has not received a lot of emphasis in the past is technology innovation. Because of this, there is limited information on how one of these key stakeholders—industry—views environmental research and technology innovation. This report summarizes information about the following:

- How research-intensive companies are rethinking investments in environmental technologies; where these companies are likely to invest, where they will not invest, and where opportunities for public-private sector partnerships are; and

- What federal policies the case-study companies would like to see to promote investments in environmental research and technology.

The information contained in this report is a synthesis of the literature on research and development (R&D) management, environmental R&D, and technology innovation—integrated with a series of interviews with senior environmental research and technology and environmental, health, and safety personnel in four research-intensive companies. It should be useful for federal, state, local, and tribal environmental and R&D policymakers and scientists; industrial managers and planners; and university researchers.

Originally created by Congress in 1991 as the Critical Technologies Institute and renamed in 1998, the Science and Technology Policy Institute is a federally funded research and development center sponsored by the National Science

Foundation and managed by RAND. The institute's mission is to help improve public policy by conducting objective, independent research and analysis on policy issues that involve science and technology. To this end, the institute

- supports the Office of Science and Technology Policy and other Executive Branch agencies, offices, and councils

- helps science and technology decisionmakers understand the likely consequences of their decisions and choose among alternative policies

- helps improve understanding in both the public and private sectors of the ways in which science and technology can better serve national objectives.

Science and Technology Policy Institute research focuses on problems of science and technology policy that involve multiple agencies. In carrying out its mission the institute consults broadly with representatives from private industry, institutions of higher education, and other nonprofit institutions.

This report is also available through RAND's web site. Inquiries regarding the Science and Technology Policy Institute or this document may be directed to:

Bruce Don
Director, Science and Technology Policy Institute
RAND
1333 H Street, N.W.
Washington, D.C. 20005
Phone: (202) 296-5000
Web: http://www.rand.org/centers/stpi/
Email: stpi@rand.org

CONTENTS

FIGURES

TABLES

> In the United States, it takes 12.2 acres to supply the average person's basic needs; in the Netherlands, 8 acres; in India, 1 acre. . . . [I]f the entire world lived like North Americans, it would take *three planet Earths* to support the *present* world population. (Emphasis added.)
>
> —*Donella Meadows (1996).*

BACKGROUND

The quotation above underlines the need to maintain economic growth without increasing—and preferably decreasing—the material and energy resources needed to achieve that growth. New technologies are needed that allow development without increased demand for those resources. In addition, new cost-competitive technologies to realize environmental benefits will help our industries remain competitive in the global marketplace.

Environmental technologies are those that "advance sustainable development by reducing risk [of human health or environmental harm], enhancing cost effectiveness [of achieving a level of environmental protection], improving process efficiency, and creating environmentally beneficial or benign products and processes." (NSTC, 1994, p. 9.)[1] While these technologies are not sufficient in themselves to achieve economic growth and improved quality of life without using more energy and material resources, they are necessary because the improvements will not occur without them.

Unfortunately, past federal environmental policy did not emphasize technological innovation as a way to achieve better environmental performance at a lower cost (OTA, 1994; EPA, 1991, 1992, and 1993). Neither was environmental policy created to accommodate differences in firms' behaviors toward environ-

---

[1]Brackets are added clarifications. Sustainable development refers to the need to allow growth while balancing economic, environmental, and social needs now and in the future.

mental issues—one policy fit all (Rejeski, 1997). As a result of this simpli-
fication, the government does not know much about how a key environmental
stakeholder—industry—views environmental research and technology innovation.[2] This lack of information is regrettable, because industry not only provides the goods and services that have an enormous effect on the environment
but it also is a font of talent, accesses even more, and outspends the federal
government $2 to $1 for R&D.

PURPOSE OF THIS REPORT

This report attempts to fill in part of the information gap and increase policymakers' understanding of these issues by illuminating *emerging* environmental
technology R&D trends in a limited number of industrial sectors. It addresses
how research-intensive companies are rethinking investments in environmental technologies; where these companies are likely to invest, where they will
not invest, and where the opportunities for public-private sector partnerships
are; and what federal policies the case-study companies would like to see to
promote investments in environmental research and technology.

HOW WE WENT ABOUT THE RESEARCH

We selected four companies identified as being among the leaders in quality
R&D processes and the treatment of environmental issues: DuPont, Monsanto,
Intel, and Xerox. These companies represent the chemicals, biotechnology, and
electronics sectors.[3] We wanted industries experiencing different rates of
change and confronting different kinds of environmental challenges. In terms
of companies, we wanted large, multinational manufacturing organizations
with significant R&D investments to focus specifically on users of environmental technologies for whom technological innovation is clearly important.
Furthermore, we searched for companies regarded as among the leaders in
their industries, hypothesizing that technologies pursued by industry leaders
would give the best insight into emerging trends.

---

[2]The list of potential stakeholders includes a diverse set of individuals and organizations—local citizens and community groups, consumers, environmental groups, industry, individual companies,
shareholders, all levels of government and tribes, etc. In turn, each of these will have various perspectives on environmental risk, priorities, costs and benefits, etc.

[3]There is no clear consensus on companies considered leaders in R&D management or environmental management. While we feel these companies are among the leaders and suitable for study,
others may disagree. These sectors were chosen somewhat arbitrarily based on interest from an ad
hoc working group of government personnel and a qualitative assessment that the technological
maturity and environmental issues among these sectors were sufficiently different from one
another.

CROSS-CUTTING LESSONS LEARNED FROM THE CASE STUDIES

The first series of questions that this study addresses deals with how research-intensive companies are rethinking investments in environmental technologies—specifically, where these companies are likely to invest, where they will not invest, and where the opportunities for public-private partnerships are.

Where the Companies Are Likely to Invest

The case-study companies are most likely to invest in technologies that improve yield, reduce emissions, increase product and process resource efficiency, create more environmentally benign products, and meet customer product requirements. They recognize that environmental issues touch a majority of their R&D investments, although they are not generally the primary reason for the investment. The innovation process they describe emphasizes the "demand pull" of technology to meet customer needs, community concerns, market trends, regulations, or their corporate environmental goals. Because environmental goals are only one of many corporate objectives that influence the R&D portfolio, leading companies integrate these issues into corporate strategic planning to improve their visibility.

Where the Companies Are Not Likely to Invest

Industries expressed less interest in remediation, monitoring, and control technologies. While often necessary to meet regulatory requirements, these technology areas are less likely to meet other corporate objectives, and, as a result, the case-study companies sought to limit the amount invested in these technologies and invested only when necessary to comply with regulations (and when outside sources are not available). The companies hoped that the pollution prevention orientation of the technologies of interest (such as yield improvement technologies and emissions reduction technologies) would to some extent lessen future requirements for remediation and control technologies.

The Role of Public-Private Partnerships

All four case-study companies were interested in public-private partnerships. However, the companies' views were mixed regarding priorities for the focus of research partnerships. Some were interested in remediation and end-of-pipe pollution control technologies, because these are precompetitive technologies.[4] For others, remediation and control technologies were no longer a priority and they were interested in collaborating on recycling and remanufacturing, yield improvement, energy efficiency, or emissions reduction technologies. Two companies specifically mentioned that the activities of public-private partnerships were used for noncritical technology enhancements or alternative technological approaches. This suggests that targeted public-private partnerships can be useful to industry if used judiciously.

LESSONS LEARNED FROM SPECIFIC FIRMS

The case studies are rich stories of how companies treat environmental technology investments. Highlights follow below.

Dupont: Looking to Solve Its Customers' Environmental Problems. Environmental technology R&D planning at DuPont has moved away from a cost containment perspective to one that focuses on market-driven opportunities and sustainability by anticipating and solving customers' environmental needs and providing products that meet environmental standards or are considered environmentally preferable. DuPont classifies its investments in environmental technologies into sustainable or environmentally preferable products and services; yield improvement, which includes co-product development and zero-waste technology; reuse and recycle; and control and abatement. Approximately 95 percent of all DuPont R&D investments have at least a modest environmental aspect to them; 65 percent have a large environmental aspect; and 15 percent are exclusively environmental.[5]

Intel: Prevention to Enable Rapid Change. Intel is known for introducing rapid series of incremental product improvements. As a consequence, product and process design occurs simultaneously and, every two years, production facilities are completely retooled. Intel invests in pollution-prevention technologies with the goal of reducing emissions below those that require an environmental permit. This approach will save the company money by avoiding the administrative expenses of obtaining permits. The approach also avoids the loss of opportunities to generate revenue because of delays to manufacturing

---

[4]Antitrust legislation prohibits firms from engaging in collaborative research on competitive technologies.

[5]Exclusive investments include control and remediation technologies; design, project engineering, and installation for abatement equipment; development of Freon® alternatives, and product and process changes for the Montreal Protocol. Largely related investments are those that improve quality or resource efficiency. For example, development and engineering for process improvements that improve first pass yield. Modestly related are those activities that might improve product quality with some minor environmental improvement.

process changes. Intel also invests in process technologies to improve water conservation, for both ultrapure and waste water because of community concerns. Other technology areas include chemical use reduction and solid-waste conservation.

Monsanto: Substituting Information for Material and Energy Resources.
Monsanto has reinvented itself as a biotechnology firm. Such a radical change takes a long time. Monsanto began investing in biology-based product research in the early 1970s. At that time, chemistry was king at Monsanto, and research focused on chemical solutions to crop management and process improvements. To transform itself, the company turned to university scientists and acquired or allied with small biotechnology companies, seed companies, pharmaceutical companies, and others to build the expertise and capability required to develop a host of biotechnology-based products in agriculture, pharmaceuticals, and animal health. It seeks to "do more with less," which is described as adding value without using more material and energy resources.

Xerox: Toward an Industrial Ecology, Closing Material and Energy Flows.
Product-related technologies in the areas of energy efficiency, chemical and physical emissions, natural resource conservation, and waste management receive approximately equal amounts of investment at Xerox. Customer requirements and Xerox's own asset-recycle-management initiative, which remanufactures and reuses copiers and other equipment, directly link these environmental investments to the bottom line. Xerox estimates show the company is avoiding *hundreds of millions of dollars* in cost because of its environmental initiatives. Because equipment is reused, savings in virgin material and energy are also achieved.[6]

CASE-STUDY COMPANIES' RECOMMENDATIONS FOR GOVERNMENT

The case-study firms see clear roles for the federal government in environmental technology R&D, and they offered a number of recommendations. The industrial sector is much more diverse than the sample represented by the companies involved in this study, and these federal policy recommendations reflect the perspective of a subset of only one stakeholder, industry. As such they might not be appropriate for federal action when all stakeholder interests are taken into account. Nevertheless, the case-study companies' rec-

---

[6]Some of the savings in material and energy used to produce virgin material and parts will be offset by the material and energy required to transport, sort, clean, and refurbish returned products. A detailed life-cycle analysis of the environmental impacts of remanufacturing is not available.

ommendations provide insight into the preferences of an important stakeholder in environmental technology policy.

Recommendations Common to Several Companies

The following themes—provide leadership, invest in science and technology,
develop markets, and protect intellectual property—came up repeatedly.

Provide Leadership. All the companies interviewed wanted the federal government to provide leadership on national environmental technology priorities.
Improved information based on scientific knowledge will aid the companies'
decisionmaking. As companies go beyond compliance, because of cost savings
or developing markets, they will face decisions about product features and
content, technological options, and emissions trades, among others.[7] Federal
leadership on national environmental priorities, operationalizing sustainability
and systems thinking, consensus building using science, and data collection
can improve industrial environmental decisionmaking and encourage the requisite R&D investments.

Invest in Science and Technology. The companies overwhelmingly felt that
continued support of a strong science and technology base was an important
role for the federal government. All the companies have ties to academia, and it
is clear from the case studies that access to university-based research helps the
companies with their own research agendas. A few mentioned the national
laboratories as sources for specialized skills or facilities.

Develop Markets. Since the companies' first priority is to meet customer needs
or requirements, policies that create "market pull" are a straightforward way to
draw environmental investment. A couple of ideas on how to create markets for
environmental products to attract additional investment, such as affirmative
procurement and labeling environmentally benign products and processes,
were suggested.

Protect Intellectual Property Rights. As industry substitutes information for
tangible resources, the protection of intellectual property carries even greater
importance to companies investing in R&D. It looks to government to be
aggressive in defending U.S. intellectual property in the global marketplace.

---

[7]The authors wish to note that all firms might not choose to move beyond the requirements established by environmental regulations. Just how widespread this behavior may be is beyond the
scope of this study. Others discuss the various strategies firms may take toward environmental
issues ranging from noncompliance to leadership. For example, see Roome (1994) and Chatterji
(1993).

Recommendations Mentioned Less Frequently

The following recommendations either were mentioned by only one company or did not receive the emphasis during the interviews that earlier topics did. These include improving regulatory policy to allow regional approaches and setting of priorities, clarifying performance-based criteria across media (air, land, water), and encouraging pollution prevention. Other suggestions included increasing funding to regulatory agencies (EPA, FDA, USDA) to attract and maintain high-quality staff, to ensure public confidence in the regulatory process, and to bring new scientific discoveries into the regulatory process;[8] developing environmental and technology policies that help companies operate in a global economy; using federal investments to improve monitoring technologies and measurement standards; and funding science and technology for sustainable products, such as economically viable methods to collect, sort, clean, and disassemble materials at the molecular level to be able to make high-quality recycled material.

DISCUSSION AND OBSERVATIONS

There is a change in corporate perspectives toward investing in environmental technologies, but not all firms will share this view. The case studies offered that these investments are "large," but only one quantitative estimate of "large" was provided. In the end, we found no systematically collected quantitative data on industrial investments in environmental technology R&D. These case studies are a limited attempt to address this gap. Yet, systematically collected data on environmental technology R&D will improve policymaking. Without information on how much industrial R&D investment has an environmental component, where these investments are being made, and how these investments are changing over time (in response to markets and policies), only a limited understanding of the effectiveness of public policies can be gleaned.

What We Heard About: The Innovation Process

Leading companies are investing in research and technology to improve the resource efficiency of their products and manufacturing processes because it is cost-effective to do so. Since much of this research deals with proprietary knowledge about products and processes, extensive collaboration is less likely than for other technology areas. Organizational and services innovations (included in the definition of environmental technology), such as closer link-

---

[8]There may be other ways to achieve these goals besides increasing funding to these agencies, such as reallocating the distribution, for example.

ages with suppliers and customers, are occurring and these will help to lessen environmental effects and may spark additional technology innovations. Technology diffusion may increase as well.

The companies relied on a rich science and technology base for environmental technology innovations. All of the companies relied on universities for new knowledge and to provide a trained workforce. Smaller technology-based companies were used for niche capabilities to complement in-house research.

Innovation, especially radical innovation, can take a long time—decades or more. Determining the right time to invest involves predictions about markets, technologies, sociopolitical conditions, and regulations. Better information and leadership on priorities can reduce some of this uncertainty. Because time horizons are long, the benefits of policies to promote environmental technology investments may not be realized for a substantial period.

What We Heard About: Policy Options to Promote Investments in Environmental Technologies

Regulations clearly influenced these firms in a variety of ways. The Toxic Release Inventory may have stimulated these companies to look at emissions as opportunities to save money; the time and expense of gaining environmental permits are causing some to practice pollution prevention; and emissions controls, hazardous-waste management, and other regulations are stimulating them to rethink their own and their customers' material and energy flows. From the companies' point of view, public confidence in the regulatory process is as important as the scientific practices employed and will improve public acceptance of new technologies and new approaches. Well-managed and scientifically rigorous environmental regulatory practices are important policy tools to negotiate the risk and uncertainty associated with new environmental technologies and thereby speed their diffusion. In spite of our observation that regulations do influence firm behavior, Intel, which is heavily involved in Project XL and other regulatory experiments, was the only company to mention any regulatory reinvention experiments at the federal or state level. In contrast, much more discussion revolved around sustainability and design-for-environment to stimulate investments and innovation.

Signals about long-term environmental priorities could be clearer. Right now the strongest signals to these firms regarding global, national, and local environmental priorities are regulations and customer preferences. More information and data will give industry and its customers the opportunity to make the right choices. Much of this information will be provided from investments in the appropriate sciences but will also require well-managed institutions and organizational structures to incorporate this knowledge into regulations, decision tools, and product criteria. Since so much of the information needed to guide decisionmaking lies beyond the scope of an individual company or an industry and requires open, vetted standards, government has a clear role here.

The case-study companies would like to see the federal government work internationally to ensure global enforcement of patent laws to protect intellectual property. Strong intellectual property rights may draw additional R&D investment, but they could also slow diffusion by limiting access to a new environmental technology or by raising its price. Since rapid diffusion is desired for environmental technologies because of the public-good aspect, new systems for protecting intellectual property must balance these somewhat competing issues.

The landscape of environmental technology R&D is complex, and no single tool will sufficiently foster a full range of environmental technology R&D investments. Federal investments in science improve the knowledge base for environmental priority-setting, and stakeholder processes will help create consensus. Support for university-based research may foster dramatically new technological options as well as train the next generation of industry-based researchers and engineers. Raising consumer awareness will increase demand for products that have improved environmental performance and help with environmental priority-setting. Public-private partnerships may help leverage funds to address common technology issues or may be effective means to build consensus among stakeholders. Each of these policy tools addresses different elements of the innovation process. Many are mutually reinforcing.

The cooperation of the following individuals helped make this study possible. They gave of their time and ideas and provided thoughtful insights that enriched this report. Although the report attempts to reflect the conversations with participating companies accurately, any errors are the responsibility of the authors.

We greatly appreciate the time the following individuals spent describing their activities: John Carberry, DuPont's Director of Environmental Technology; Terry McManus, Intel's Assembly-Test Manufacturing Environmental Health and Safety Manager; Tim Mohin, Intel's Manager of Corporate Environmental Affairs; from Monsanto, Phil Brodsky, Vice President, Corporate Research and Environmental Technology, Earl Beaver, Director, Waste Minimization, and Laurence O'Neill, Director, Environmental Communications; and from Xerox, Allan E. Dugan, Senior Vice President, Corporate Strategic Services, Mark Myers, Senior Vice President, Corporate Research and Technology, Rafik Loutfy, Vice President, Strategy and Innovation for Corporate Research and Technology, Jack Azar, Director, Environment, Health, and Safety, Jim Cleveland, Manager, Quality, Technology and Engineering, Ralph Sholts, Manager, Recycle Technology Development, Stephen Dunn, Manager, Environmental Technology, Patricia Calkins, Manager, Environmental Products and Technology, Ronald Hess, Manager, Environmental Engineering Programs/Operations, and Joseph Stulb, Manager, Environmental Engineering Operations.

David Rejeski, Council on Environmental Quality (formerly of OSTP), and Kelly Kirkpatrick of OSTP have both been extremely supportive and engaged clients throughout the entire research process. Lloyd Dixon and Ron Hess both gave thoughtful, timely technical reviews that have improved the report's quality. In addition, we would like to thank Cynthia Cook for her informal review of the entire document and Nicole DeHoratius, Elisa Eiseman, and Deborah Sole for their reviews of the Intel, Monsanto, and DuPont cases, respectively. And Daniel Sheehan and Jerry Sollinger have contributed immensely to the report's presentation. The contents and conclusions are the sole responsibility of the authors.

API	American Petroleum Institute
BAT	Best available technology
CAAA	Clean Air Act Amendments of 1990
CFC	Chlorofluorocarbon
CIETP	Chemical Industry Environmental Technology Partnership
CRADA	Cooperative research and development agreement
CSI	Common Sense Initiative
CWRT	Center for Waste Reduction Technologies
DNA	Deoxyribonucleic acid
DoC	Department of Commerce
DoE	Department of Energy
DRAM	Dynamic Random Access Memory (chip)
EH&S	Environmental, health, and safety
EIRMA	European Industrial Research Management Association
ELP	Environmental Leadership Program
EPA	Environmental Protection Agency
EPRI	Electric Power Research Institute
FACA	Federal Advisory Committee Act
FDA	Food and Drug Administration
GEMI	Global Environmental Management Initiative
HAPs	Hazardous air pollutants
ICOLP	International Cooperative for Ozone Layer Protection
IRI	Industrial Research Institute
ISO	Industrial Standardization Organization
MOU	Memorandum of understanding
NGO	Nongovernmental organization

NIH	National Institutes of Health
NRTEE	National Roundtable on the Environment and Economy
NSF	National Science Foundation
NSTC	National Science and Technology Council
ODS	Ozone-depleting substances
OTA	Office of Technology Assessment
PACE	Product and cycle-time excellence
PCSD	President's Council on Sustainable Development
PERF	Petroleum Environmental Research Fund
PFCs	Perfluorocarbons
PNGV	Partnership for New-Generation Vehicle
Project XL	Project eXcellence and Leadership
PRP	Potentially responsible party
R&D	Research and development
RCRA	Resource Conservation and Recovery Act
RTDF	Remediation Technologies Development Forum
S&T	Science and technology
SARA	Superfund Amendments and Reauthorization Act
SIC	Standard Industrial Code
STEP	Strategies for Today's Environmental Partnerships
TDS	Total dissolved solids
TQM	Total quality management
TRI	Toxics Release Inventory
UPW	Ultrapure water
USCAR	U.S. Council for Automotive Research
USDA	U.S. Department of Agriculture
VOC	Volatile organic compound
WAVE	Water Alliance for Voluntary Efficiency
WBCSD	World Business Council for Sustainable Development

Chapter One

INTRODUCTION

BACKGROUND

Investments in environmental technologies are important because sustainability is not likely without technological innovation[1] and new cost-competitive techniques to realize environmental benefits will help our industries remain competitive in the global marketplace. The concept of sustainability is important. The world's population is increasing; its material and energy resource base is not. If the United States is to continue its growth and other nations are to enjoy their own as well, new techniques are needed that provide development without increased demand for those resources. Although considerable debate swirls around the precise meaning of sustainability, its broad definition is generally accepted as economic growth without increasing—and preferably decreasing—the material and energy resources needed to achieve that growth.[2] While sustainability is entering industry's lexicon—many corporate environmental, health, and safety reports use the term; the World Business Council for Sustainable Development (WBSCD) is sharing ideas about sustainable business concepts;[3] and all the companies interviewed for this study use the term—translating it into day-to-day operations is still problematic because the concept remains relatively ill-defined and ambiguous.

Environmental technologies are those that "advance sustainable development by reducing risk [of harm to human health or the environment], enhancing cost effectiveness [of achieving a level of environmental protection], improving

---

[1]More than 30 years ago environmentalists Paul Ehrlich and Barry Commoner related the human environmental burden to population, wealth, and technology (Hart, 1997, p. 70).

[2]This definition was developed by the World Commission on Environment and Development, *Our Common Future*, 1987, p. 46; often referred to as the Bruntland Report. For a more extensive discussion on the terminology, see Schmidheiny (1992); Perrings (1994); and Fussler with James (1996).

[3]The WBCSD is an international coalition of companies (Fussler with James, 1996, p. 129).

process efficiency, and creating environmentally beneficial or benign products and processes. "Technology" includes hardware, software, systems, and services." (NSTC, 1994, p. 9.) Four categories of environmental technologies—avoidance, control, monitoring and assessment, and remediation—have been defined for the purposes of policy development (NSTC, 1994, pp. 42–43). (See Appendix A for the full definitions of these technology categories.) Whereas technological innovation alone cannot completely resolve problems of environmental performance, the introduction of chlorofluorocarbons (CFCs) substitutes and numerous energy-efficient technologies, among other developments, demonstrate how innovation has allowed economic activity to grow while reducing undesirable environmental effects.

Private industry invests in environmental technology research and development (R&D). It implements or deploys new environmental technologies. And the products and services that it designs and produces have large environmental impacts. Information is not available for environmental technologies specifically but information on *overall* R&D investment suggests that where federal R&D investment is an important policy tool, the magnitude of these investments pales in comparison to the private sector. Preliminary data for 1997 show that out of a total of nearly $206 billion in R&D investment, private industry funded an estimated $133.3 billion while the federal government funded $62.7 billion (64.8 and 30.5 percent, respectively). The industrial sector is both the largest source of R&D investments and the leading R&D performing sector, surpassing the federal government in such investments since 1980.[4] Federal environmental technology strategy, presented in *Bridge to a Sustainable Future* (NSTC, 1995), recognized the importance of the industrial sector and sought to improve public- and private-sector relationships to stimulate the development of environmental technologies and to increase their diffusion. Yet, unfortunately federal policy is developed with little systematically collected information on what industry is doing.

STUDY OBJECTIVES

To better structure its environmental R&D priorities and policies, the federal government must develop a clear understanding of how research-intensive industries are rethinking investments in environmental technologies; altering their environmental management processes; where these industries are likely to invest, where they will not invest, and where there are opportunities for public-private sector partnerships; and how existing and future competitive pressures are likely to affect industrial environmental R&D strategies.

---

[4]The remaining $9.7 billion of R&D was funded by nonprofit organizations (NSF, 1997).

This report seeks to illuminate *emerging* environmental R&D trends in a limited number of industrial sectors and to increase government decisionmakers' understanding of private-sector R&D activities and processes. Areas of particular interest are the following:

- motivations for investments in new environmental technologies

- amount of R&D resources devoted to environmental technologies

- role of public-private collaboration and cooperation in environmental technology development

- case-study companies' recommendations for federal policies to encourage or assist environmental technology development and diffusion.

Ultimately, this information will help improve federal R&D policies to promote environmental technologies. Federal policies considered in the past include direct federal investment, public-private research partnerships, federal support for high-risk technology demonstrations, use of federal laboratories' capabilities and resources, research and experimentation tax credits, and federal-state collaboration on research efforts. Other policies that influence environmental technology R&D investments include environmental regulations, product liability laws, green labeling programs, federal and state procurement criteria, foreign aid and technology assistance, education and training investments, and programs to collect and disseminate environmental information.[5]

THE STUDY APPROACH: LITERATURE REVIEWS AND CASE STUDIES

We took a two-step approach. In one step, we reviewed and synthesized the literature on R&D management, environmental R&D, and technology innovation. This process provided an essential context for the second step, which involved case studies of leading firms in different industries. Because the relevant R&D cannot be fully understood outside of the corporate context, the case studies give the reader a more complete picture than would the literature review alone of the challenges and opportunities for environmental R&D that each participating company faces. However, the methodology provides only a small window on a very large world, and, consequently, conclusions cannot be carried too far. The literature review tempers this limitation by providing some information about the generalizability of the lessons learned from the case studies.

---

[5]This is not an exhaustive list of potential federal policies, but it does provide an illustrative overview. For more specifics on federal policy options see NSTC (1994) or OTA (1994).

Structured interviews were performed with company representatives largely from R&D and environmental, health, and safety organizations. Follow-up conversations were performed with interviewees to clarify points or collect missing information. On average, interviews for each company lasted between one and two days in total. The interviews were augmented with information from corporate annual reports, corporate environmental reports, press releases, and articles in the open literature on environmental and R&D management. Originally we had hoped to assess the content of the case studies with quantitative information from the open literature. Since we found this information limited, our ability to critically assess what we heard in the interviews was also limited.

Company Selection Criteria

We employed several criteria to select industry sectors and firms. We sought industries with different rates of technological advancement and different environmental issues, hypothesizing that different pressures in these realms will lead to different policy recommendations. For firms, we wanted large, multinational manufacturing companies with significant R&D investments who are users of environmental technologies and for whom technological innovation is clearly important. These companies are the "environmental problem owners," or regulated industries, as distinguished from the generally smaller firms that provide services and technologies to address environmental problems.[6]

To illuminate *emerging* trends, companies identified as leaders in quality R&D processes and the treatment of environmental issues were selected. By employing these selection criteria, we seek to draw attention to policies that can aid the innovators among large, multinational companies. The major limitation of these specific selection criteria is that the results cannot be generalized to all of industry.

Industries Selected

We selected the chemical, electronics, and biotechnology industries for study.[7] The chemical sector is a diverse industrial sector with more than 70,000 different chemical products registered and more than 9,000 corporations in the

---

[6]A companion study of the generally smaller, environmental technology providers was performed by the Environmental Law Institute (ELI) (ELI, 1997).

[7]There is no clear consensus on companies considered leaders in R&D management or environmental management. While we feel these companies are among the leaders and suitable for study, others may disagree. These sectors were chosen somewhat arbitrarily based on interest from an ad hoc working group of government personnel and a subjective assessment that the technological maturity and environmental issues were sufficiently variable.

standard industrial code classification (Lenz and LaFrance, 1996). Because of this, the chemical industry has often been characterized by its tiers: commodity, intermediate, and specialty. The most basic tier includes commodity chemicals, such as chlorine, benzene, and xylene, which are inputs to other chemical processes. The commodities tier is generally considered technologically mature. The intermediate chemicals, such as polyester films or fibers, begin with commodity chemicals to create chemicals used to manufacture other products, such as electronics, automobiles, furniture, and appliances. Specialty chemicals are often final products, such as paints, fertilizers, and fibers. These products as well as their manufacture can be toxic or hazardous to human health and the environment. Environmental regulatory policy has addressed the chemicals industry, among others, which spends a significant amount in capital and operations expenditures to comply.[8] Environmental regulations cover materials-handling in general; emissions of toxic or hazardous materials to the air, water, land; and remediation of previous disposal sites. The chemical industry as a whole consumes a large amount of energy, using 5.95 quadrillion Btus for feedstocks and power in 1996 (more than 6 percent of total U.S. energy consumption and nearly 17 percent of total U.S. industrial energy consumption) (U.S. DoE, 1998).[9]

The semiconductor area of the electronics industry has experienced rapid innovation. For example, the density of transistors per chip has improved between 1971 and 1998 at an exponential rate (some refer to this constant doubling as Moore's Law) (Hutcheson and Hutcheson, 1996). In contrast to the chemicals industry, electronics manufacturing was initially thought to be a "clean" industry. Environmental issues for the semiconductor industry include those that are regulated—air and water emissions of acids and solvents—and those that potentially may be regulated or are of regional concern—water consumption, solid waste generation and disposal, and heavy metals dispersion (EPA, 1995).

Studies of the biotechnology sector estimate there to be approximately 1,300, predominately small (fewer than 135 employees), firms (Paugh and LaFrance, 1997, pp. 28–29). R&D spending in this sector has been estimated at 77 to 83 percent of sales (in 1991 and 1992) (Fusfeld, 1994, p. 277) or 36 percent of all expenses in 1995 (Paugh and LaFrance, 1997, p. 30), making it the most research-intensive industrial sector (outside of the defense sector) in the country. The biotechnology sector is also experiencing rapid technological change (Paugh and LaFrance, 1997, p. 30). Like the chemical sector, it encompasses a

---

[8]See U.S. Bureau of the Census (1994) for the exact figures.

[9]In 1996, the entire industrial sector consumed 35.5 quadrillion Btus, the residential and commercial sector 33.7 quadrillion Btus, and the transportation sector 24.7 quadrillion Btus for a total of 93.9 (Energy Information Agency, 1998).

broad range of products. These products cut across several standard industrial classification codes, including pharmaceuticals, plants, medical services, pesticides, and bioremediation microbes. Environmental issues for this industry are complex, reflecting the breadth and immaturity of the technology. Predominant concerns outside of the medical subsector are how these products either degrade (for example pesticides) or spread (for example transgenic plants) in the environment. As an example, concern over transgenic crops effects on biodiversity and human health is a major issue for this sector. Yet, on the other hand, these very same products may have environmental benefits (by lowering overall pesticide use or reducing water use). The subsector of bioremediation microbes is used to render hazardous dump sites safe.

Companies Selected

We chose DuPont, Xerox, Intel, and Monsanto to study. We looked for companies with reputations for innovation and high-quality management in both environmental and R&D processes. This was determined in part by identifying firms that were active in voluntary programs (such as the Environmental Protection Agency's [EPA's] Project eXcellence and Leadership [Project XL], EPA's 33/50 Initiative, EPA's Water Alliance for Voluntary Efficiency [WAVE], and the Department of Energy's Industries of the Future Program) or management initiatives (such as Global Environmental Management Initiative [GEMI], the Chemical Manufacturers' Responsible Care®, American Petroleum Institute's Strategies for Today's Environmental Partnership [STEP], and the WBCSD). In addition we looked for award winners (environmental and management awards), Toxic Release Inventory (TRI) emissions, amount of R&D investment, mentions in the literature discussing high quality or innovative firms, and reputation. In the end, we found limited information was readily available in the open literature that would allow us to quantitatively identify industry leaders both within an industrial sector and across sectors. (For the detailed supporting data gathered to inform the selection criteria please see Appendix B.) These companies—DuPont, Intel, Monsanto, and Xerox—were selected based on subjective judgment, using a combination of the limited information available and the recommendations of select knowledgeable industry and government personnel.

Information on company size and R&D investments is shown in Table 1.1.

The sample selected does not represent a cross-section of industry, but it attempts to capture the diversity of industry, at least among industry leaders. Each company interviewed has some level of experience with a range of federal programs—notably the EPA's Remediation Technologies Development Forum

Table 1.1

Characteristics of Participating Corporations

Company	1997 Sales or Revenues (Billions)	1997 R&D Investment (Millions)	R&D Intensity[a]	Sector
DuPont	45.1	1,116	2.4[b]	Chemicals
Intel	25.1	2,347	9.4	Electronics/ Semiconductors
Monsanto	7.5	902	12.0	Biotechnology/ Life sciences
Xerox	18.2	1,079	5.9	Electronics/ Office equipment

[a]R&D intensity is the R&D investment divided by sales multiplied by 100.

[b]DuPont's R&D intensity would rise to an estimated 3.9 percent without Conoco oil and gas division, which is in the process of being sold (assuming its R&D investments follow the petroleum industry's average of 0.8 percent of revenues).

RTDF), the DoE's Industries of the Future Program, the Department of Commerce's Advanced Technology Partnerships, various cooperative R&D agreements with federal laboratories, and the EPA's regulatory reinvention Project XL.

REPORT ROADMAP

The remainder of the report is organized as follows. First, to provide a larger context for the case studies so that the details of the cases are as meaningful as possible, Chapter Two provides a very brief overview of the literature on innovation theories, the major trends in R&D, environmental technology R&D investments, and public-private partnerships. Innovation theories provide a useful framework for thinking through how policy might influence the creation and diffusion of environmental technologies. Readers familiar with innovation theory, R&D trends including environmental technologies, or public-private partnerships need not read these subsections. Chapter Three presents emerging themes in environmental technologies investments based on the case studies. The participating companies were asked to recommend federal policies to promote and encourage investments in environmental R&D. Their recommendations are provided in Chapter Four. Chapter Five presents some final observations on the case studies and the implications for policy.

The case studies in their entirety are rich summaries of how the four companies treat environmental innovation. DuPont seeks to mitigate its customers' environmental problems, to develop products to meet environmental standards, and to improve the yields of its own processes (Appendix C). Intel is constrained by the environmental permitting process and therefore seeks technologies that reduce emissions (Appendix D). Monsanto has embraced sustainability as the strategic thrust of its business enterprise and is investing in radical technology innovation (Appendix E). Xerox, with its asset-recycle management program, continues to introduce incremental technological improvements that reduce overall material usage (Appendix F).

The DuPont, Intel, Monsanto, and Xerox stories illustrate how industry leaders treat environmental investments. Where federal action is warranted, this information should help policymakers structure federal R&D investments and develop policies that effectively encourage continuous innovation for improved industrial environmental performance.

Chapter Two

WHAT THE LITERATURE SAYS ABOUT
INNOVATION, R&D, AND PARTNERSHIPS

The collective understanding of how technological innovation occurs is constantly changing, and much has been learned recently about the process that challenges old notions of how policy should encourage innovation.[1] Because past environmental policy did not much concern itself with encouraging constant technological innovation (OTA, 1994; EPA, 1991, 1992, and 1993), a brief overview of this process is provided here as useful background material. It is also important, before delving into the case studies, to define and summarize some key terms and concepts. This material is based on the review of the literature. Information from the case studies will be presented in later sections.

INVENTION, INNOVATION, AND DIFFUSION

The innovation process is commonly divided into three stages: invention, innovation, and diffusion. Invention is the initial conception of an idea. Innovation is the first application of the idea to actual practice by a firm or a consumer. Diffusion is the process by which additional firms and consumers adopt the new technology.[2] These distinctions highlight innovation as more than having a good idea—and recognize that market success and market penetration provide rewards.

In the past, people viewed the entire innovation process as a linear process. First came the invention (idea), then the innovation (market application), and finally the diffusion (market adoption). More recent research shows the process

---

[1]There is a large literature of innovation theories. For an overview of the three major paradigms in the literature see Sundbo (1995).

[2]Often referred to as the Schumpeterian trilogy after Joseph Schumpeter, one of the first economists to link innovation to economic growth. Note that innovation is often used to refer to the entire process of invention, innovation, and diffusion or to invention and innovation only. For the purposes of this report, innovation refers to the entire process. Other terms will be used only to add emphasis.

rarely occurs in such a straightforward manner. Rather, experts have noted the genesis for a new invention can occur at any point in the cycle, one invention often leads to another, and one invention may lead to multiple innovations.[3] Moreover, successful innovation and diffusion may depend on the presence of complementary technologies, sociopolitical conditions, and regulations (Schneiderman, 1991, p. 56; Myers and Rosenbloom, 1996). Experts have also suggested that the linear model is also outdated in that it implies that all companies pursue every innovation in the same manner (Utterback, 1994, p. 79; Kline, undated)—a view that is entirely too simplistic.

Another important insight from the innovation literature is the importance of differences among those who adopt a novel technology (during diffusion). Typically the diffusion, or market adoption, of new technologies over time follows an S-shaped curve in the aggregate (called an epidemic or logistics curve).[4] This process has been known to take as long as 50 years, although rates of five to 25 years have been observed more often (Stoneman, 1987, p. 51). Several overarching themes have been used to explain rate and patterns of diffusion. Researchers have described it in the context of information and learning (or becoming aware of the new technology's capabilities), differences in potential adopters (such as capital infrastructure or cost structures), and changes in these differences among potential adopters as the market expands (as a technology becomes more widespread it may become more useful to a broader group, e.g., fax machines). Some have pointed out that government policy has emphasized information and learning and paid little attention to adopter heterogeneity as one of the underlying drivers in diffusion (Stoneman, 1987, p. 51).

A new technology almost always has higher cost, less reliability, and fewer features than the version it will eventually become. And, as a result, it is often hard for an innovation to dislodge established technologies. "Niche markets" play an important role in helping new technologies develop to the point at which they can take on a wider market (Bower and Christensen, 1995). The photovoltaic energy industry in the United States is often considered to be using this as a strategy. Photovoltaic cells are used in niche markets—i.e., in remote telecommunications and irrigation. As the technology develops and operational experience is gained, potentially larger markets may be sought.

---

[3]For more on innovation in general see Stoneman (1987), Utterback (1994), or Dodgson and Rothwell (1994).

[4]The S-shaped curve called the epidemic or logistics curve assumes a homogeneous population whose members have an equal probability of coming into contact with each other (Feder and Umali, 1993). This curve is a simplified model of diffusion. Other models, such as the Probit model, Bayesian learning models, and game theoretic models, have been used to describe diffusion. See Lissoni and Metcalfe (1994) for a brief review.

Product and Process Innovation and Industry Cycles

Another distinction made in the literature is between product innovation, which is a new product, and process innovation, which is a new way to make products. The two kinds of innovations are highly interactive, and environmental technology can be of both types. For instance, a new energy-efficient light bulb would be a product innovation. In semiconductor manufacturing, aqueous cleaning to replace the use of solvents containing CFCs would be a process innovation. Within a new industry, the initial emphasis will be on product innovations.[5]

Over time, researchers found that companies compete on the combinations of product features that are offered until the market is established and a dominant set of features develops. Then the focus turns to process innovations as the competitive basis. At this time, product innovations still appear, but they focus on the dominant features desired by the established market. Once a dominant product design is established, the number of companies serving the market usually decreases, and the barriers to entry increase. The organizational structure and management focus will differ depending on which phase an industry occupies.[6] Industries or product classes typically make incremental improvements leaving the organization of an industry intact (i.e., market leaders remain market leaders) until a radical innovation appears.[7] Radical innovation, on the other hand, dramatically changes features offered by a technology. These innovations are most often introduced by companies new to the market, and they generally produce a new set of industry leaders.[8] Some of the environmental literature suggests that radical innovation is required to move toward sustainability (Fussler with James, 1996).

Implications for Environmental Technology Policy

What does this all mean for environmental technology policy? If environmental policy seeks to encourage innovation, several themes emerge from the literature.

- Innovation is a complex process with multiple feedback loops that involve the creation of new technologies (invention), the development of markets

---

[5]This discussion is relevant to a product class as well.

[6]For more on this topic see Utterback (1994, pp. 79–92).

[7]Incremental improvements are those that modestly improve the features of the dominant design or lower its cost of manufacture.

[8]For more on this topic see Utterback (1994, pp. 145–165); Green, McMeekin, and Irwin (1994); Fussler with James (1996, pp. 8–12); or Kline (undated, p. 43).

(innovation), and the spread of new technologies across diverse companies and industries (diffusion). As a result, policies that focus on R&D alone may not be as effective as considering the entire process. More than likely, no single policy tool will be sufficient to promote technology development and deployment.

- There is variability among firms and between industries. Policies will be more effective if they explicitly recognize that firms do not all have the same cost-performance curves and the same capability to gather and incorporate new information. To remain effective, policies may have to change as industries mature and their emphasis moves away from large technological leaps in product technologies and toward incremental process improvements.

- Technology innovation and diffusion can take decades during which great uncertainty exists regarding the technology development and the market. As a result, the benefits generated by federal policies might not be realized for a long time.

These are important points because in the past environmental regulation envisioned that all firms would use the same technology. We now recognize the importance of experimenting with different approaches and of providing the grounds for new technologies to develop and deploy.

R&D TRENDS

R&D is important to innovation and diffusion because this process provides new inventions. Recent experience with budget constraints and market trends has changed the way R&D is viewed and managed. Industrial R&D in particular is following several overall trends. Because little is known explicitly about environmental R&D investments and because they are likely to follow the same trends as other R&D investments, the major trends are summarized below.

R&D Investments Are More Closely Linked to Business Goals

The decline in industrial R&D during the early part of the 1990s caused companies to more closely evaluate the effectiveness of these investments and to look for opportunities to leverage funds. Trends to leverage more limited R&D resources include greater emphasis on outsourcing R&D, acquiring R&D either through licensing or buying a smaller firm, entering into private-private alliances, consortiums, or joint ventures and participating in public-private partnerships (NSF, 1996; Council on Competitiveness, 1996). In addition there is a greater emphasis on linking corporate R&D projects to the needs of operating or production divisions. The amount of corporate R&D directly funded by the operating divisions is increasing. As a result, there is more pressure to respond to market-driven needs or shorter-term focused projects and less to long-term, more basic research. Those still performing these activities are often pressured into "directed" basic research.[9]

Reducing Time-to-Market Has Become Paramount in Managing R&D

There is also a premium for getting to market early; it has been estimated that a product that is delivered to market three months early could earn extra profits up to five times the total cost of the R&D for that product (Institute for the Future, 1995, p. 12; Council on Competitiveness, 1996, pp. 16–18). One way to shorten the time-to-market is to enter into an alliance with a firm that has technology or market access that complements in-house capabilities. Alliances among firms have become important for companies that seek to rapidly extend capability to respond to market opportunities. "Trading" in intellectual property to gain market advantage worldwide through acquisition, licensing, etc., is another way to take a new concept to market rapidly (Myers and Rosenbloom, 1996; Fusfeld, 1994).

Markets and Technologies Are Global

Corporate R&D has become globalized in response to expanding markets worldwide. More and more frequently, firms will perform R&D to respond to special requirements (i.e., different countries' regulations and customer interests) of international markets for products sold in various regions of the world. Facilities near key markets seek to incorporate local market knowledge more rapidly. "And technology development will increasingly occur outside the United States; thus, cooperative means to monitor, develop, or acquire technology will become increasingly important." (NSF, 1996; Council on Competitiveness, 1996.) The United Nations has estimated that 36,000 transnational corporations operate a network of 175,000 foreign affiliates worldwide and international investment has blurred national boundaries, causing a high degree of global economic integration."(Florida, 1995, p. 50.)

---

[9]However, several of the company representatives interviewed for this study—Xerox, Monsanto, and DuPont—said they have not reduced their investments in basic research, although no quantitative information was provided to support this claim. They did note that among scientific disciplines priorities were changing (a couple said toward information and biological sciences).

Information on Environmental Technology Investments Is Sparse

Not surprisingly, little is known about the amount of investment in environmental technologies. Because universally accepted definitions of environmental technologies do not exist, estimates of these investments are quite variable. An Industrial Research Institute (IRI) survey of members performed in 1991 found that respondents were spending approximately 13 percent of their R&D investments on environmental technologies. If we apply this percentage to the 1991 NSF estimate for total industrial R&D investments, an estimate of approximately $10 billion was invested by industry for environmental technology R&D in that year.[10] However, the author of the IRI study cautions that rigorous definitions were not imposed and that the sample was not completely representative of all of industry (Rushton, 1993, pp. 13–21; Rushton, 1997). Moreover, since the IRI survey was performed around the time of Clean Air Act Amendments of 1990, the number from a survey today may be lower. An Office of Technology Assessment (OTA) study reviewed the IRI study along with NSF data and select data from the petroleum and pulp and paper industries. OTA reports that a more likely estimate for investments in *pollution-control technologies*, only one component of all environmental technologies, is between 1 percent and 2 percent of overall R&D funds.[11] Another study, performed by the European Industrial Research Management Association (EIRMA) in 1992 took a different look at environmental R&D. The results of this study reported that nearly half of all R&D projects have a *significant environment and safety component* (Rich, 1993, pp. 16–23). While performed in similar time periods, each of these studies had different definitions of environmental technologies and, as a result, came to variable conclusions regarding the amount industry invests in environmental technologies.

PUBLIC-PRIVATE RESEARCH PARTNERSHIPS

Why do companies collaborate? What role could public-private partnering play to promote sound environmental technologies and what would be the likely payoff? Because public-private partnerships have received much attention in stakeholder events and are policy tools of great interest to federal policymakers, a discussion of these issues follows.

Partnerships can take many forms, including

---

[10]This percentage of 1997's industrial investments is more than $17 billion.

[11]This figure refers to the regulated industries, which OTA estimates support half of all US investments in pollution-control technology R&D (OTA, 1994, pp. 308–310).

cooperative arrangements engaging companies, universities, and government agencies and laboratories in varying combinations to pool resources in pursuit of a shared R&D objective. Such partnerships vary widely in scale and scope, from company-to-company tie-ups to complex networks involving all three of the stakeholders in the innovation process. The common thread that runs throughout all partnerships is the joint commitment of participants—whether firms, universities, or federal entities—to share costs, resources, and experiences and to draw strength from each other by leveraging capabilities. (Council on Competitiveness, 1996, Foreword.)

Why Does Industry Collaborate on Research?

The simple fact is that industry is not in business to cooperate. It is not an easy matter to have five or ten companies agree on technical programs of common interest sufficiently far removed from the proprietary base of their product lines. The objectives of such common programs usually emphasize basic research or generic technology, and those interests constitute a small percentage of industrial research budgets. (Fusfeld, 1994, p. 308.)

In the past, companies have been hesitant to enter into cooperative arrangements either with other industry members or with government for a host of reasons: potential loss of proprietary or highly competitive technology, loss of decisionmaking power and control, fear of antitrust litigation, and perception of low benefits (Link and Bauer, 1989, p. 14). However, legislative changes to loosen restrictions on collaborating, combined with market realities, have made collaboration a part of day-to-day business. The number of joint research partnerships has grown to 450 since 1984. In the 1990s, an average of 60 research partnerships were established each year.[12]

Two of the primary motivations for collaborative partnerships discussed in the literature are the desire (1) to share expenditures on R&D, particularly when available resources are limited or when benefits are difficult to capture for an individual firm, and (2) to bring together diverse technical capabilities and market knowledge when developing complex or innovative products and services. In more and more instances, an individual company cannot afford to see a broad-based research agenda through commercialization and market diffusion without the benefits of partnerships. Financial limits and the pressures of

---

[12]The National Cooperative Research Act of 1984 requires firms interested in establishing joint research partnerships—a new organization created by two or more organizations to conduct R&D—to register with the Department of Justice to avoid potential antitrust litigation (NSF, 1996, pp. 4-16 to 4-18).

time have increasingly led companies to engage in research partnerships and alliances.[13]

An Overview of Select Environmental Technology Research Partnerships

Environmental technology research partnerships are unique because of the social benefits that these technologies generate. Because of these benefits and the environmental issues in common with government operations, public involvement may be warranted. In addition, much of the technology deployment literature suggests that government investments alone may not be as effective as those made in partnership with industry. These are also the kinds of technologies one would like to see diffused as rapidly as possible, and partnerships may be one mechanism to speed this process.

Industry publications have expressed favorable views on partnerships and in particular those formed to develop environmental technologies (Rushton, 1993; Rich, 1993). From a company's point of view, collaborative research on environmental technologies may be an opportunity to share expenses for technologies necessary to comply with environmental regulations. They may also be a way to reduce the risks associated with introducing new technologies to comply with regulations and the risks of environmental liability.

Many research partnerships have addressed environmental issues shared by government and industry. The following examples—International Cooperative for Ozone Layer Protection (ICOLP), RTDF, Chemical Industry Environmental Technology Partnership (CIETP), Petroleum Environmental Research Fund (PERF)—illustrate the range of research partnerships established to address environmental issues in terms of organizational structure and the kinds of problems they address.

---

[13]For example, in 1990, IBM and Siemens agreed to joint development of an advanced dynamic random access memory chip (DRAM). Although either of these two companies could have developed the new-generation DRAM individually, working together improved their chances of getting to the market earlier rather than later, which is very important in the electronics industry. In addition, if IBM or Siemens had pursued DRAMs individually, other R&D programs in their portfolios might have been delayed and other product lines jeopardized. Later, in 1992, IBM and Siemens drew from their earlier experiences together and collaborated with Toshiba to develop a 256-megabit DRAM. Their aim was "to share the exploding costs"—$1 billion to develop the chip and $1 billion to build a plant. IBM brought extensive memory chip design expertise to the partnership; Siemens provided access to the European market; and Toshiba brought leading manufacturing techniques. Even in this instance, these companies could have gone it alone, but another important benefit to collaborative efforts is the ease with which a company expands its options—either in terms of markets, technological approaches, manufacturing capability, or manufacturing capacity (Fusfeld, 1994, p. 99 and pp. 103–104).

ICOLP, a nonprofit partnership, is one of the earlier examples of a collaborative effort that had government involvement but that was primarily led by industry. In 1989, AT&T and the EPA's Global Change Working Group spearheaded ICOLP, an alliance of corporate and government CFC users, to develop alternatives to the ozone-depleting CFCs (note, ICOLP's name has since changed).[14] By allying with Boeing, Digital, Ford, General Electric, Honeywell, Motorola, Northern Telecom, Texas Instruments, and the U.S. Air Force, "problem-holders" successfully used their combined market power, technological expertise, and financial resources to find CFC alternatives (Piasecki, 1994, pp. 75–95). A Ford employee stated that "ICOLP showed us technical capabilities that would have cost us multimillions of dollars to develop internally." (ICOLP, 1996, p. 30.) Moreover, by finding alternatives sooner rather than later, companies avoided costs associated with CFC price increases, EPA taxes, and CFC emissions standards (Piasecki, 1994, pp. 75–95). Because cooperation on the elimination of ozone-depleting solvents CFC-113 and methyl chloroform was so successful, ICOLP continues today to develop, exchange, and promote innovative solutions to environmental issues primarily through information exchange and evaluation of technological solutions.

RTDF was established in 1992 by the EPA to foster collaborative research, development, demonstration, and testing and evaluation between the public and private sectors for remediation technologies to address mutual cleanup problems. According to the RTDF literature, the participants have brought various strengths and assets to the collaboration. The EPA facilitates the operation of the RTDF Steering Committee and contributes its research efforts to the jointly led projects. The EPA also works with the states and other regulatory agencies to conduct demonstration projects. Industrial participants help set priorities based on remediation problems they face and provide both in-kind and monetary resources to support joint projects. The U.S. Department of Energy (DoE), U.S. Department of Defense (DoD), and other federal agencies suggest priority areas for their remediation problems and contribute funds to research projects. DoE and DoD also provide contaminated facilities as needed for field-scale testing. Universities and other research institutions provide science and engineering expertise and help ensure that engineering and scientific principles are followed. As of 1997, the RTDF was supporting approximately $22 million in R&D activities. Approximately 44 percent of funding for RTDF comes from industry, 55 percent from government, and 1 percent from

---

[14]ICOLP has changed its name from the Industry Cooperative for Ozone Layer Protection (ICOLP) to the International Cooperative for Ozone Layer Protection (ICOLP) to the International Cooperative for Environmental Leadership (ICEL). It has numerous projects covering water conservation, solvents use, global warming, and ozone layer protection. Its homepage can be found at http://www.icel.org.

academia. Action teams develop consensus on priority work areas. Self-managed teams implement the projects by sharing information, developing project plans, and implementing these plans. The results are then disseminated to encourage acceptance of the technologies (EPA, 1997). One case-study interviewee stated that this partnership is particularly effective because all stakeholders—industry, government, academia, and nongovernmental organizations (NGOs)—are engaged, and each brings an important perspective to the partnership. Industry's role is to focus the technical program; government provides the technology and the test opportunities and shepherds the legal issues through the system; academia provides creative ideas; and the NGOs and government ensure that public interests are represented (to facilitate acceptance).[15]

Other examples of industry research partnerships that cover environmental technologies are the chemical industry's CIETP and the petroleum refining industry's PERF. According to one of the CIETP founders, CIETP is a for-profit, limited-liability entity that seeks to develop environmental technologies for the chemical industry. CIETP was formed in 1995 by Battelle, Air Products and Chemicals, Inc.; Akzo Nobel Chemicals, Inc.; and E.I. duPont de Nemours and Company. The Council for Chemical Research is an advisory member, and the American Chemical Society is a supporter. The aim in setting up this partnership is to share expenses and risks for environmental technology developments that will help interested parties in the chemical industry meet corporate sustainability goals, address environmental regulations, or reduce potential liabilities more cost-effectively. Technologies developed within this partnership are owned by participating companies and licensed to others (Sciance, 1994 and 1996). According to John Carberry of DuPont, CIETP, as opposed to the RTDF, is a more appropriate forum to develop specific treatment and abatement technologies because of concerns over technology liability. CIETP is also more appropriate for broader chemical-processing technologies where competitive issues can become an issue (Carberry, 1997).

The petroleum industry's PERF is a more informal forum for petroleum companies and others to collaborate on environmental technology and to discuss, develop, and fund cooperative ventures. The organization is a venue for discussing potential projects. Contracts are established on a project basis, and project funding primarily takes the form of in-kind work. The projects under

---

[15]Remediation technology and service providers, as opposed to the problem holders, are not a part of this forum. There are at least a couple of reasons for this. One is that the forum seeks to address all potential solutions without bias toward a particular approach. Once potential solutions or approaches are identified, then providers may be brought in to contribute advice and expertise. If the forum develops a new technology, it may then be licensed to these providers. Another reason is that the smaller remediation technology and service providers invest little in R&D (Carberry, 1997).

way during 1996–1998 focused on air experiments to inform risk-assessment methodologies and bioremediation technologies. Research project results are published in the public domain after an embargo period of approximately six months. Since pollution-prevention technologies are proprietary and plant-specific, they are not addressed by PERF (Krewinghaus, 1997).

The Partnership for a New Generation of Vehicles (PNGV) and the National Center for Manufacturing Sciences are two public-private partnerships that develop environmental technologies. Established in 1993, PNGV is a research consortium of the federal agencies and laboratories; universities; the research teams of Chrysler, Ford, and General Motors; and the major suppliers. It was created to produce radically more fuel-efficient automobiles (White, 1997, p. 68). It also supports research on improved manufacturing technologies and efficient conventional automobiles. The National Center for Manufacturing Sciences was established to provide a legal and administrative framework for developing improved manufacturing processes. Environmentally conscious manufacturing is one of the research areas. Member companies come from many industrial sectors—autos, electronics, jet engines, machine tools, health care, and others. Member companies believe that every dollar they spend on this partnership leverages an additional five dollars of research.[16]

These examples show that environmental research partnerships have been developed in the past for a variety of purposes including the following:

- Address important environmental issues that may not receive enough R&D funding within one company because they are not necessarily market-driven.

- Eliminate duplication among problem-holders.

- Reduce the time it takes to develop new technology.

- Avert regulation.

- Encourage radical technological approaches.

- Engage regulators and other stakeholders to facilitate technology acceptance.

- Speed the diffusion of technology.

No one knows the amount of industrial resources that could, or *should*, be available for collaborative efforts for environmental technologies. Very little

---

[16]http://www.ncms.org, accessed 1997.

information could be found to define the issues. What could be collected is presented next.

How Much Industrial Research Can Be Leveraged with Public-Private Partnerships?

One industry expert has conjectured that if the amount of precompetitive or generic research is analogous to research typically performed at central R&D facilities of large corporations, then historical industrial spending patterns suggest that potentially up to 10 percent of industrial research investments overall could be available for R&D partnerships of some nature.[17] Proportionally, smaller firms would be expected to have a larger share of their R&D portfolio devoted to collaborative efforts because they have limited funds and expertise to support a wide-ranging research agenda. Data for the semiconductor industry, where collaboration is especially prominent, show this pattern (Table 2.1). They also show that a small percentage of research funding, in the range of 4 to 6 percent, goes to research consortiums.

The U.S. Council for Automotive Research (USCAR)/PNGV, the consortium among Chrysler, Ford, and General Motors, their suppliers, and several federal government agencies provides another data point. A 1992 estimate suggeststhat $300 million per year is being invested in PNGV by the U.S.

Table 2.1

Distribution of Collaborative Research by Firm Size in the Semiconductor Industry During 1990 (percentage of R&D investment)

Alliance Type	Company Size		
	Less than $100 million	$100 million to $1 billion	More than $1 billion
Total technical collaborations	41	16	9
Company-to-company technical collaborations	37	10	5
Technical consortiums	4	6	4

SOURCE: Fusfeld, 1994, p. 105.

---

[17]Antitrust legislation prohibits firms from engaging in collaborative research on competitive technologies (Fusfeld, 1994, pp. 114–115). Fusfeld recommends that no more than 20 percent of the total industrial research investments be devoted to collaborative efforts. Another important point made by the author is that "the ability and willingness of member companies to exploit" the results of collaborative efforts is another critical, but completely separate, matter.

government. This funding is matched by the three automakers. The industry portion amounts to less than 3 percent of the more than $11 billion invested by the big three automakers in 1992.[18] Another data point is the utility industry—a regulated industry that is undergoing major restructuring—so these values may be uncharacteristic of industry in general. An estimate of utility industry investment in R&D is $523 million (FY97$) for 1996. In that year, its major collaborative research organization, the Electric Power Research Institute (EPRI), was funded at $472 million (FY97$) ($311 million from member dues alone), amounting to approximately 60 to 90 percent of the total research investment in that year.[19] These data are summarized in Table 2.2.

The only data we could readily find on environmental R&D collaboration were for PERF and EPRI. PERF had projects totaling approximately $10 million to $15 million per year of in-kind work during 1996–1997, significantly higher than previous years. This collaboration alone accounts for less than 0.5 percent of

Table 2.2

Summary of Research Consortiums

Industry Sector	Percent of R&D Funds in Consortiums (Consortium name)	Year
Corporate laboratory research	5 to 10	1994[a]
Semiconductor industry	4 to 6	1990
Automobile industry	Less than 3 (PNGV)	1992
Electric utility industry	60 to 90 (EPRI)	1996

[a]A time frame was not offered for these data. The source was published in 1994.

the petroleum industry R&D expenditures. The other organization in the petroleum industry that does some collaborative research, only some of which

---

[18]Fusfeld (1994, p. 309) states that USCAR is investing this amount, but his description of the collaboration sounds like it is a PNGV investment (Chapman, 1997).

[19]Another research consortium relevant to the electric utility industry is the Gas Research Institute funded at $179 million (FY97$) in 1996. While some of its research may be relevant to the electric utilities, it receives funding from other sectors, such as industrial consumers, whose R&D investments are not included in the utility industry R&D estimate. Therefore, this consortium was not included in our estimate. Also note that R&D investments by the utility industry have been higher in the past. In 1993, it invested $778 million (FY97$) into R&D. EPRI funding in 1993 was approximately $620 million (FY97$) overall. This is 80 percent of all the R&D funding in the utilities industry (President's Committee, 1997, pp. 2-12–2-13).

is environmental, is the American Petroleum Institute (API), which is funded at less than $15 million per year (Krewinghaus, 1997). At most, this would double the amount of funding devoted to collaborative research on environmental matters in the petroleum industry to 1 percent of all R&D. Estimates for EPRI show that 12 percent of its research focuses on environmental issues.[20] If EPRI is somewhere between 60 and 90 percent of all electric industry R&D, then environmental research performed in this consortium amounts to somewhere between 7 to 11 percent of the total.

In sum, very limited data suggest that partnerships for all kinds of technologies may be 3 to 10 percent of R&D and in unique circumstances as high as 90 percent. Considering that environmental R&D is a small subset of overall R&D, the percentage available for research partnerships is smaller yet. Even though these percentages are small, at the lowest end, 0.5 percent of $133.3 billion in industry-funded R&D in 1997 is still more than $650 million that could be leveraged with industry to work on environmental technologies each year. If focused on priority areas, this is a significant amount of research. However, dollar values present only part of the picture. One of the major advantages of partnerships is that they are effective mechanisms for bringing diverse technological capabilities together quickly to find solutions to problems. Public-private partnerships can improve information flow between government and industry, and they can, if managed well, eliminate duplication of effort, facilitate technology acceptance, and speed technology diffusion.[21]

---

[20]The other EPRI research areas and the percentage of funding were electricity end use (21 percent), nuclear power (21 percent), generation (19 percent), power delivery (19 percent), and strategic technology R&D (8 percent) (President's Committee, 1997, pp. 2–12).

[21]The authors note that we found much of the partnership literature to be descriptive rather than critical.

Chapter Three

CASE-STUDY LESSONS

Several lessons have emerged from the case studies of leading companies. We have found that industry leaders view environmental technology R&D broadly and they link the environment to a large portion of all R&D. Environmental technology investments are largely driven by corporate strategy, which seeks to meet market demands and achieve greater resource efficiency in products, services, and manufacturing processes as well as to comply with environmental regulations. These changes are implemented in part through effective partnering with suppliers and customers. The companies interviewed are multinational and as such respond to global market trends with technology from global sources. These similarities across the case-study companies are attended by differences in environmental concerns and technology needs. The information provided in this chapter is based on the case studies unless explicitly noted otherwise.

HOW THE FIRMS VIEW "ENVIRONMENTAL TECHNOLOGY"

The case-study companies characterized environmental technology R&D investments similarly. Most, but not all, indicated that the portion of R&D driven solely by environmental concerns was quite small. Indeed, the term "environmental technology" most often connotes a narrow set of technologies—those necessary to meet regulatory or product requirements. However, they also recognized that a large portion of R&D investments had an environmental component. These are R&D investments made for other purposes, such as cost reduction or product enhancements, and are not *driven* by environmental concerns. Yet these technologies may still have modest to large environmental impacts. For example, DuPont estimated that 15 percent of its R&D portfolio is directly related to environmental needs; 50 percent has a large environmental component; 30 percent has a modest environmental element; and only 5 percent is not at all related to environmental issues.[1]

At Xerox, environmental, health, and safety issues are integral to R&D (a large portion of which is product-oriented). New materials undergo health and safety analyses. Xerox personnel felt that environmentally preferable practices are becoming a source of competitive advantage. Therefore, these issues affect most of Xerox's R&D. However, a relatively small part of the R&D investments, on the order of 1 to 2 percent, are driven by environmental goals exclusive of global safety, regulatory, and environmental product standards.

Neither Intel nor Monsanto identified the amount of R&D dollars spent on these technologies, although Monsanto suggested that the amount spent on technologies required for environmental regulations was quite small, perhaps less than 5 percent. Intel stated that it integrates environmental concerns in all decisionmaking and therefore does not identify these costs separately. The terms "environmental technology," "control," "avoidance," "monitoring and assessment," and "remediation" are not uniformly used by the firms interviewed.

Therefore, while environmental issues are not the primary driver to much of the case-study companies' R&D investment, it is explicitly recognized that a majority of these investments have an environmental component. For example, cost-reduction investments may increase resource efficiency or reduce waste and new product offerings may have environmentally preferable features. The case-study companies seek to exploit these opportunities to their competitive advantage.

FIRMS EMPHASIZE CUSTOMER PRIORITIES AND RESOURCE EFFICIENCY

The dominant R&D priorities for all the companies interviewed are meeting customer interests and improving resource efficiency (both product and process). The companies invest in these environmental technologies because it is cost-effective to do so. Both incremental and radical technological approaches are used to accomplish these goals.

---

[1]Exclusive investments include control and remediation technologies; design, project engineering, and installation for abatement equipment; development of Freon® alternatives; and product and process changes for the Montreal Protocol. Largely related investments are those that improve quality or resource efficiency. For example, development and engineering for process improvements that improve first pass yield. Modestly related are those activities that might improve product quality with some minor environmental improvement (Carberry, 1997).

In terms of customer interest, Xerox has identified several product features that customers care about, such as toner efficiency, energy efficiency, and noise emissions. DuPont seeks to be the supplier of choice by making products that meet certain environmental requirements (such as recycled material content). DuPont also seeks to invest in technologies that solve its customers' environmental problems (such as more-benign photoprocessing chemicals). In some cases, companies invest in R&D in anticipation of future customer needs. For example, Monsanto is assessing how water shortages around the world may create emerging markets for new product offerings.

Resource efficiency of products and processes is another area receiving much attention. For example, in 1996 DuPont estimated that the next 50 percent improvement in waste reduction per pound of product will save the company three to five billion dollars.[2] Besides investing in technologies to create environmentally improved products and services, it is investing in yield improvement technologies including zero-waste processes and reuse and recycle opportunities at the molecular level.[3] Intel, on the other hand, desires a lot of process flexibility because of its rapid rate of innovation.[4] To avoid the lengthy and expensive process of getting environmental permits, it is investing in a host of areas to reduce its emissions—particularly air emissions—to below regulatory thresholds. In response to community concerns, Intel is also investing in reuse and recycle technologies for water. Other environmental technology areas of concern for Intel include alternative manufacturing processes, chemical reuse and management technologies, natural-gas and boiler-emission reductions, and control technology development. Monsanto is interested in product technologies that minimize resource use. For example, plants that bear edible fruit and whose leaves can be used to produce biodegradable plastic. Xerox's asset-recycle management initiative, which increases resource efficiency through remanufacturing, has avoided hundreds of millions of dollars in costs, according to its own estimates. As a result, reductions in the generation of solid waste have been large and there should be upstream savings in energy, material, and air and water emissions because virgin material use is much lower.[5]

---

[2] The largest element of this estimate was the amount of revenues waste material could have generated if it were product. Waste per pound of product will be reduced with the range of DuPont's environmental technology investments—yield improvements, sustainable products, reuse/recycle, zero emissions, and co-product development technologies (DuPont, 1996b, quoting Paul Tebo, Vice President Safety, Health, and Environment).

[3] DuPont is also investing in reduced emissions, treatment, and abatement technologies.

[4] Because plant equipment is modified frequently, Intel has opportunities to incorporate new environmental technologies and revise the manufacturing process.

[5] Some of the savings in material and energy used to produce virgin material and parts will be offset by the material and energy required to transport, sort, clean, and refurbish returned products. A detailed life-cycle analysis of the environmental impacts of remanufacturing is not available.

Technologies that contribute to customer priorities and resource efficiency of products and processes are areas in which environmental considerations align well with the primary corporate competitiveness and profit-making objectives. Two specific technology areas that were frequently mentioned were biotechnology and information technology. Several interviewees felt that these technologies in particular could be employed to add value or create products and services without using more material and energy resources in potentially more efficient and environmentally benign manufacturing processes.

R&D PLANNING LINKS ENVIRONMENTAL TECHNOLOGY TO BUSINESS OBJECTIVES

Another observation is that all the companies interviewed are to some degree integrating environmental issues into their strategic and R&D planning processes. Since many corporate objectives influence the R&D portfolio, environmental technologies can receive priority and funding if they explicitly relate the benefits of investment to the bottom line—historically these companies had used environmental regulations to justify environmental technology investments. Now that they have a broader view of environmental technology the companies actively integrate environmental considerations into R&D planning and link these considerations to corporate strategy and profit. The next several paragraphs provide three examples from our case studies.

Since 1988, DuPont has gradually begun to realize that environmental R&D investments could be used to meet strategic business objectives other than cost containment. Now DuPont seeks to become the supplier of choice because of a positive environmental image in the marketplace; provide products that meet environmental requirements; and solve the customers' environmental problems for them. It has incorporated these and other environmental objectives into its R&D and other business planning processes. For example, every 6 to 12 months the vice president of each business area meets with the chairman of DuPont. At these meetings, the customers' environmental problems and the business unit's methods for taking advantage of them are major discussion items (along with plant emissions and product stewardship). Evaluating customers' wants and needs is the primary objective of this process, which is performed within the R&D organization as well. For example, the vice president of environmental technology meets with each business area's technical director and laboratory personnel to discuss their programs, using a template for sustainable business developed by the WBCSD and DuPont's own Safety, Health, and Environmental Vision (see Appendix C for the template and the vision statement). This planning process is especially important in light of the business units' possession of the lion's share of R&D funds. Finally, an approach employed by DuPont to improve the relevance of its research is called product and cycle-time excellence (PACE). This is a practice to evaluate research programs against objectives for commercialization—financial, market position, capital efficiency, product quality, and other objectives—early in the research process before large sums of money are spent.

Monsanto elevated sustainability to a strategic business issue after a 1995 meeting called Global Forum. Its chief executive officer, Robert Shapiro, has said that businesses that help raise living standards of the world's population—such as products that ultimately improve health, food availability, energy supply, and water supply—will in turn help create economic growth, improve quality of life, and better the environment. Monsanto personnel also recognize that improvements in standards of living will in turn increase the demand for its products—the obvious priority for stockholders. Moreover, global stability will make doing business easier. Sustainability is operationalized at Monsanto as a key process of doing more with less. At its core, sustainability is finding opportunities created by fostering growth without increasing the use of material and energy.[6] To integrate sustainability thinking into business planning processes Monsanto is developing decision tools—such as product sustainability criteria and life-cycle assessment techniques, eco-efficiency[7] assessments of process material and energy flows, full cost-accounting techniques, and so on. Underlying this business strategy is a reliance on new technology, specifically biotechnology.

At Xerox a product-oriented environmental technology plan and research needs report was developed by its environmental, health, and safety (EH&S) group in 1995 (it was called *der Gedanken* team). This plan, along with customer and regulatory trends data, is currently being marketed to the groups that drive R&D investments—the product-design units and research laboratories. Ultimately, Xerox EH&S personnel hope to integrate completely the treatment of environmental attributes into the product-design units' technology planning. They estimate that they are about two years from this goal.

The case studies illustrate that, for these firms, the connection between the environment and other business goals (such as profitability, cost reduction, market access) has been made. Concern about market demands worldwide—specifically product features, public concerns, environmental regulations, and competitor actions—influence environmental investments. Competitiveness issues, as opposed to environmental regulations alone, are thus stimulating

---

[6]See Monsanto's 1997 Report on Sustainable Development including Environmental, Safety and Health Performance, March 1998, p. 2.

[7]Eco-efficiency is defined in Fiksel (1996, p. 499) as, "The ability of a managed entity to simultaneously meet cost, quality, and performance goals, reduce environmental impacts, and conserve valuable resources."

environmental technology investments.[8] The anticipatory nature of strategy means that determining the right time for investments becomes a key question for companies. Although environmental regulations are still important, competitiveness concerns offer great incentive for innovation. To the extent policy recognizes these drivers and enhances them, more investment may be generated.

OPPORTUNITIES ARE LINKED TO CHANGES ALONG THE VALUE-CHAIN

> We need to provide future generations with ways to meet their needs in a way that requires what [Monsanto CEO Robert] Shapiro calls less "stuff." I state it a little differently. We need to provide the value our technology and products offer but with less through-put of energy and raw materials and little or no waste at the other end. And we have to devise ways of working with suppliers, customers and consumers to make sure that our products do not end up in dumps after they are no longer useful.
>
> —*Jack Krol, DuPont Chairman of the Board and former President and CEO (Krol, 1997).*[9]

Greater integration and cooperation on environmental issues is now apparent between the manufacturers and suppliers, distributors, and customers. As mentioned earlier, DuPont is using its core capability of handling hazardous chemicals to solve its customers' environmental problems so that they can comply more easily with environmental regulations and internal corporate goals. For example, it is working with Intel and others to develop a process so that Intel can collect waste perfluoromethane and send it back to DuPont for reuse. Intel is cooperating with other suppliers to reduce emissions of global warming gases, collect waste chemicals for reuse, and improve ultrapure water recovery. Intel anticipates that by working with equipment suppliers to develop new wet benches that more efficiently clean silicon wafers, ultrapure water use will be reduced by 40 percent. This in turn will reduce total water demand by 300,000 gallons per day at each facility (Intel, 1996b, p. 12). Xerox is working with suppliers to recycle and remanufacture components, to incorporate recycled materials, and to use recycled paper. It is working with customers and

---

[8]While regulations have been shown to stimulate some innovation, environmental regulations for a variety of reasons can also have the unintended consequence of hindering innovation. Uncertainty regarding the form of the regulation, its enforcement, the administrative burdens of verifying performance and modifying permits, market segmentation resulting from differing standards, and the liabilities associated with potential performance failure can prevent long-term innovation.

[9]In this speech, Krol outlined three challenges for the chemical industry—value creation, technology, and sustainability.

distributors to increase the amount of equipment and components that are returned for remanufacturing or recycling. Monsanto is educating farmers on no-till farming techniques, which are feasible with the combination herbicide and plant products that Monsanto sells.[10] It is a member of the Center for Waste Reduction Technologies (CWRT), which includes companies from the chemical, petroleum, pharmaceutical, and upstream and downstream manufacturing industries, collaborating to identify, develop, and transfer environmental technologies.

From an environmental standpoint, this close connectivity along the value-chain could mean greater improvements as complete product systems are optimized for material and energy content. It could also mean that technology innovation and diffusion will occur more rapidly because the tighter linkages will shorten the time it takes for new information to influence product and process changes.

IN A GLOBAL MARKET, MANY POLICIES INFLUENCE R&D

Since these companies sell to markets worldwide, they must meet unique customer requirements and various environmental standards, especially the product-related ones, in each of these markets. For example, in 1994 Xerox began to supply paper to the European market that consisted of 100 percent recycled material produced in chlorine-free processes. In that same year, it introduced paper made of 20 percent recycled material to the U.S. market. These differences reflect customer preferences and regulations. Xerox must also comply with European product standards (which ban polybrominates for example) and takeback legislation, as well as with U.S. standards (on energy efficiency for example). Trade rules, historically created to protect against dumping of used or inferior equipment in developing markets, complicate the export of remanufactured equipment around the globe because, according to these rules, this can be interpreted as the selling of waste. In the area of biotechnology, a real debate is going on in the United States and Europe about different labeling policies for food products made from genetically modified plants. In the United States, products must be labeled only if there is a potential health or safety risk, while in Europe, food products made from transgenic crops must be labeled regardless of risk, emphasizing consumers' right to know. In some cases, food products made from transgenic crops are mixed with conventional ones during processing (as in the case of soybean oil), which makes labeling, and trade, problematic. Lastly, pending European takeback legislation has caused U.S.

---

[10]No-till farming reduces energy use and soil erosion.

electronics manufacturers to develop and apply design-for-recycling and dis-assembly practices and technologies.

Global market trends also influence the products and services that the companies seek to provide. For example, Monsanto is using biotechnology to address global food and water shortages. It trains farmers in developing countries on environmentally preferable no-till farming techniques, which in turn generates demand for its herbicide and plant products. DuPont is trying to anticipate and mitigate the effects of potential changes to the availability of feedstock materials (or process inputs) globally.

The flow of technical know-how transnationally is increasing as companies form alliances with suppliers, distributors, and other firms. Multinationals especially will use technology from any source, U.S. or foreign, to address problems or market needs. In addition, their ability to introduce new technology around the globe speeds technology diffusion; particularly as companies form transnational alliances. For example, while a U.S. firm ultimately won the competition, the Monsanto Environmental Challenge grant program was advertised internationally—Monsanto wanted to maximize the creativity brought to bear against its problem.

In sum, U.S. policy is not the only policy that influences environmental technology investments. Multinationals invest in R&D to meet the environmental requirements of all relevant markets, and they drive innovation in their suppliers by informing them of market trends. Technology is fluid—U.S. technology is applied to meet overseas market needs and vice versa.

HETEROGENEITY IN THE FIRMS

Previous subsections emphasize the similarities in the companies. Yet several industrial sectors were sampled to illuminate how environmental issues and corporate responses can vary. Each case study tells us something different about industrial environmental technologies. The companies differ in the degree to which environmental issues are linked to strategy and the bottom line. Some emphasize product technologies, while others emphasize process technologies. And some companies are extending core competencies while others are creating new ones. For some companies, environmental permitting practices constrain the process change for select product lines. For others, these permits are not that important. A brief synopsis of each case illustrates this variability.

DuPont Helps Solve Its Customers' Environmental Problems

Environmental R&D planning at DuPont has moved away from a cost containment perspective to one that focuses on market-driven opportunities and sustainability. It emphasizes anticipating and solving customers' environmental needs and applying DuPont's hazardous material–handling experience and capabilities to its customers' operations. DuPont's R&D planning also emphasizes product research and technology that may make it the supplier of choice because of a reputation for environmental excellence or because its products meet environmental standards required by customers. It also seeks to anticipate potential disruptions to the availability of feedstock (process input) materials. R&D planning is structured to address customer environmental needs, sustainability issues (using the WBCSD template), and DuPont's own Safety, Health and the Environment Vision.

The classes of environmental technologies that DuPont is investing in are sustainable or environmentally preferable products and services; yield improvement, which includes co-product development and zero-waste technology; reuse and recycle; and control and abatement. For example, DuPont is developing an environmentally benign process (it requires no heavy metals, petroleum, or toxic chemicals, and the liquid effluent is biodegradable) to manufacture polyester intermediates using glucose from corn starch (a renewable) and enzymes from a microorganism created through recombinant DNA. The microbial by-product can by used as animal feed. And DuPont has a patented process to unzip these molecules for recycle (Krol, 1997; Holliday, 1998). The majority of its investments are in yield improvement and sustainable product technologies. DuPont executives have stated that biotechnology, as applied to the chemical and life sciences businesses, has the potential "to replace raw material, energy, and capital-intensive processes with low-temperature, low-pressure, nontoxic, zero-waste, and cheaper routes to products important to society." (Holliday in Reisch, 1998, pp. 21–25.) Approximately 95 percent of all DuPont R&D investments have at least a modest environmental aspect to them; 65 percent have a large environmental aspect; and 15 percent are exclusively environmental.[11]

---

[11]Exclusive investments include control and remediation technologies; design, project engineering, and installation for abatement equipment; development of Freon® alternatives, product and process changes for the Montreal Protocol, etc. Largely related investments are those that improve quality or resource efficiency. For example, development and engineering for process improvements that improve first pass yield. Modestly related are those activities that might improve product quality with some minor environmental improvement (Carberry, 1997).

Intel Practices Pollution Prevention to Enable Rapid Change

Intel is known for introducing rapid series of incremental product improvements—its motto is "quick or dead." As a consequence, product and process designs occur simultaneously. Every two years, production facilities are completely retooled. This turnover suggests that plenty of opportunities exist to incorporate new environmental technologies. Environmental regulations are constraining and costly for this pace of innovation—specifically the environmental permitting process of the Clean Air Act. Consequently, Intel is investing in pollution-prevention technologies, with the goal of reducing emissions below permit thresholds to eliminate the permitting requirement entirely. This will save the company money in permitting expenses as well as in lost opportunities to generate revenue because of delays to manufacturing process changes. Because of community priorities, Intel is also investing in process technologies to improve water conservation, for both ultrapure water and waste water.

Public comments on Intel's Project XL agreement illustrate the need for toxicological data, risk-assessment methods, and monitoring technology to determine indisputable actual environmental performance. Finally, public-private research partnerships are not well suited for this pace, either. Therefore, most research partnerships are used to provide alternatives or to refine approaches already in practice at Intel.

Monsanto Is Substituting Information for Material Resources

Monsanto has achieved radical change, and it has created an entirely new competency in biotechnology. It began its biology-based product research in the early 1970s. At that time, chemistry was king at Monsanto and research was focused on chemical solutions to crop management and process improvements. Over time, the company hired university scientists and acquired or allied with small biotechnology companies, seed companies, pharmaceutical companies, and others to gain the knowledge and capability required to develop a host of biotechnology-based products in agriculture, pharmaceuticals, and animal health. Nearly 25 years after its initial investment in biology, more than 40 years after J. D. Watson and F. H. C. Crick described deoxyribonucleic acid (DNA), and 139 years after Gregor Mendel began his experiments with peas, Monsanto has introduced several new products based on its research investments. Monsanto is applying its expertise in biotechnology to sustainability. It seeks to "do more with less," which is described as adding value without using more material and energy resources. To operationalize sustainability strategies, it is developing business analysis tools in collaboration with others to guide decisionmaking and to integrate environmental concerns into overall business planning.

Xerox Is Closing Material and Energy Flows

Environmental R&D at Xerox is primarily product-oriented, although some investment is made in process-related technology. Product-related technologies in the areas of energy efficiency, chemical and physical emissions, natural resource conservation, and waste management receive approximately equal amounts of investment. Customer requirements, environmental regulations, and Xerox's own asset-recycle-management initiative, which remanufactures and reuses equipment, directly link these environmental investments to the bottom line. The company is achieving *hundreds of millions of dollars* in cost avoidance because of asset-recycle-management. Because components and equipment are reused, savings in virgin material and energy use are also achieved. Xerox tracks worldwide trends in environmental regulations, product standards, and consumer interests and links these trends to business goals for the product units, design teams, and research laboratories.

With rapid product innovation, environmental planning for permitting is quickly becoming a constraint on process cycle times (in some cases it is already the primary constraint). Early manufacturing facility planning and pollution prevention through material selection and process design are two tactics for reducing the complexity and time required by the environmental permitting process.

Each story provides a brief overview of the focus of environmental R&D at the case-study firms, and each differs in the emphasis placed on product versus process innovation; the pace and scope of these innovations; and the specific connections that have been made among environmental investments, strategy, and competitive advantage. The complete case studies presented in the appendixes provide much richer information on how the companies treat environmental investments.

SUMMARY

Investments in environmental technology R&D appear to be small, between 1 percent and 15 percent, if only those investments exclusively driven by environmental concerns are considered. However, all the companies interviewed stated that a large portion of industrial R&D has an environmental component. And the perception of environmental technology investments is changing from those solely made to meet the requirements of environmental regulations, which are minimized, to a broader range of market-driven investments.

The innovation process described by the companies emphasizes the "demand pull" of technology—that is, the market value of these product features or process efficiencies attracts investments in R&D. Now, investments are more readily made for products or processes that link environmental issues to profitability or to customer demands. By integrating environmental technology planning into corporate strategic planning and business planning, these companies are actively seeking those technologies in which environmental and competitive assets coincide.

The multinationals were also sensitive to global trends in customer environmental needs, voluntary environmental standards, regulations, and resource scarcity. Their product offerings and technology planning processes incorporate these trends.

Chapter Four

POLICY RECOMMENDATIONS FROM
DUPONT, INTEL, MONSANTO, AND XEROX

Interviewees were asked for recommendations for federal policies that would stimulate investments in environmental technologies. The remainder of this chapter presents a summary of the four companies' recommendations (based exclusively on the interviews unless otherwise noted). Note, these recommendations for federal policy come from the perspective of only one major stakeholder, industry. And much more diversity exists in the industrial sector than is represented by the companies involved in this study. In addition, even the broad-minded companies included in this analysis necessarily have limited objectives for such policy. As a result, these priorities might not be appropriate for federal action when the interests of all stakeholder groups are considered. Nevertheless, the recommendations offered by these companies provide insight into the preferences of an important stakeholder in environmental technology policy.

The first section presents recommendations about a set of priorities for federal environmental technology policy that received the greatest emphasis with all the companies interviewed. The second section discusses public-private partnerships because they are a policy tool of special interest to the client and this administration. The last section concludes with a list of recommendations offered by only one company.

COMMON RECOMMENDATIONS EMERGED DESPITE THE DIVERSITY OF INDUSTRIAL SECTORS

The following themes—invest in science and technology, provide leadership, create markets, and protect intellectual property—were raised in nearly every interview. And they were the topics that received the most attention during the discussion of desired federal action.

Support the Scientific Infrastructure

The companies interviewed overwhelmingly felt that continued support for a strong science and technology base was an important, multidimensional role for the federal government. As described in the appendixes, all the companies had ties to academia; a few mentioned the national laboratories. It is clear from the case studies that access to university-based research helps the companies with their research agendas by either augmenting internal research or gaining unique expertise that did not exist within the company. Sometimes university-based research was used to monitor scientific progress in a particular field. The national laboratories were recognized in some interviews for their specialized technological skills and facilities, while one company suggested that they had a unique capability for demonstration projects, especially concepts generated at the universities.

The specific recommendations for supporting the scientific infrastructure were the following:

- Support a strong science and technology infrastructure, particularly university-based research to ensure that a scientifically skilled workforce is available. Invest in research in the basic sciences (one company specifically mentioned the applied sciences) so the United States continues to be a font of knowledge generation. One interviewee suggested doubling the investment in federal R&D because he thinks that we are on the cusp of a new technological wave. Another company's representative suggested using the National Institutes of Health (NIH) model of university support. Scientific and technology areas mentioned include biotechnology, chemistry based on biological analogies, information technology, nanotechnology, energy efficiency, and monitoring technologies. Other priorities were to maintain specialized skills or unique facilities and equipment at the federal laboratory that may be shared by industry and government.

- Fund science and technology to enable development of sustainable products. Companies provided two specific examples. One, develop economically viable ways to collect, sort, clean, and disassemble materials at the molecular level to make recycled material. Two, understand and develop low-cost ways to reduce the amount of energy required to address other environmental issues; for example, develop inexpensive, lower–energy intensity alternatives to recycling or zero-emissions manufacturing processes.

- Improve public education on scientific and environmental matters so that scientific discussions may be more easily understood, new technologies may be more readily accepted, environmentally preferable products may be distinguished and accepted, and communities may become better equipped to establish their environmental goals and priorities.

One final point made by a couple of companies was to develop mechanisms to incorporate new scientific knowledge into environmental and other regulatory policies. One company even suggested increasing funding to such regulatory agencies as the EPA, the Food and Drug Administration (FDA), and the U.S. Department of Agriculture (USDA) to improve the ability to attract and maintain high-quality staff, to ensure public confidence in the regulatory process, and to facilitate the introduction of new scientific discoveries into the regulatory process.[1]

Provide Leadership and Information

All the companies interviewed suggested that the federal government provide information (based on scientific knowledge) to aid decisionmaking. Federal leadership on environmental priorities developed by incorporating scientific knowledge, consensus building, and data collection can improve industrial environmental decisionmaking and reduce investment uncertainty. In the past and to a large extent today, regulations provide the primary guidance to companies on their investment decisions regarding the environment. As more companies go beyond compliance, because of cost savings or developing markets, they will face decisions about product features and content, technological options, emissions trade-offs, and other elements.[2] Information from the federal government on environmental priorities would help this process. The specific recommendations for federal action were as follows:

- Lead and provide a forum with all stakeholders for environmental technology R&D priority setting to build consensus among stakeholders and provide leadership, much like the forums that addressed ozone-depleting substances. One company suggested the forum could develop a national environmental technology agenda and action plan with priorities that could be implemented in industry, universities, and federal laboratories. The difference from this and previous visioning or strategy exercises, which provided a good conceptual overview, is that the concepts would be developed more completely and operationalized into projects. Progress and priorities would also be reevaluated on a continuous basis.

---

[1] A reviewer noted that improving the quality of regulatory agencies' research staff may be achieved through means other than funding increases.

[2] Note that not all firms will choose to go beyond compliance. For some, compliance may always be their objective.

- Promote sustainability and systems-thinking among all stakeholders. The actions required to carry out this recommendation include

  — working with industry to understand what sustainability means in terms that can be made operational and incorporated into new products and services;

  — developing systems models of product systems and resource flows that combine both environmental and economic considerations;

  — collecting data and creating analysis tools for sustainable product design and development, life-cycle assessment, and full cost accounting; and

  — collecting and providing data on environmental impacts, such as those for bioaccumulating chemicals.

Develop Markets

Developing markets was not emphasized as much as investing in the science and technology infrastructure and providing information and leadership were. Yet ideas to develop demand were frequently offered. Moreover, based on the cases themselves it is clear that the companies' first priority was to meet customer needs or requirements. Thus, policies that create "market pull" may be a straightforward way to draw environmental technology investment. Creating markets is to some extent intertwined with other federal policies. For example, developing markets would be difficult without the scientific basis for determining what is environmentally preferable or without stakeholder consensus on environmental priorities. The suggestions listed below offer some ideas about how to create markets for environmental products.

- Use voluntary eco-labels such as EPA's Energy Star label (combined with consumer education) to create markets for environmentally preferable products based on scientific knowledge of what constitutes an environmentally benign activity or product.

- Direct the purchasing power of all public entities—federal, state, and local—to create markets for environmentally preferable products. For example, revise public procurement rules to allow the purchase of recycled or remanufactured products so long as performance is maintained.

- Focus on policies that reward positive behavior or promote continual improvement in environmental performance, such as EPA's green chemistry program and Wa$te Wi$e. Incorporate pollution prevention into all regulations.

Protect Investment in R&D

Research investment can generate new products or processes that provide competitive advantage for the firm. Research can also create a portfolio of marketable intellectual property (in the form of patents) that provides currency for the establishment of strategic partnerships to assure market access (Myers and Rosenbloom, 1996, pp. 14–18). Appropriability—or the ability to retain the value of R&D investments—is obviously a high priority for any company investing in R&D. As industry substitutes information for material resources the protection of intellectual property carries even greater importance. Thus, many firms suggested that the federal government work internationally to ensure global enforcement of patent laws to protect intellectual property.

PRIORITIES FOR PUBLIC-PRIVATE RESEARCH PARTNERSHIPS VARIED

All four case-study companies were involved with public-private research partnerships, and the majority were interested in continuing them. However, partnerships were not emphasized as much as other policy tools were. Two companies specifically mentioned that the activities of public-private partnerships were used for noncritical technology enhancements or alternative technological approaches. Another mentioned that federal priorities for such partnerships did not coincide with its own technology interests and as a result its project proposals had been unfunded. One suggested that the federal government has inherent difficulties in changing investment priorities, making cooperation difficult over time.

The companies' views were also mixed regarding priorities for the focus of research partnerships. Some felt that such partnerships were appropriate for remediation and end-of-pipe pollution control technologies because these are more readily generic or precompetitive technologies. For others, remediation and control technologies were no longer a priority. These firms were interested in collaborating on recycling and remanufacturing, yield improvement, energy efficiency, or emissions-reduction technologies. Specific technologies mentioned by one company were energy efficiency technologies (such as improved power supplies and improved fusing), durable plastics recycling technologies, materials-sorting technologies, product-disassembly technologies (such as degreasing agents), and product life-extension technologies.

For the most part, RTDF was considered a good partnership model by the interviewees, although one company had problems with patent infringement by the U.S. Air Force. The PNGV was also mentioned as a good partnership model but note that none of the case-study companies were involved with this partnership.[3]

OTHER RECOMMENDATIONS HIGHLIGHT THE HETEROGENEITY

The following recommendations either were mentioned by only one company or did not receive the emphasis during the interviews that previous topics did. While several recommendations related to environmental regulations, each company had its own suggestion for improvement or changes to the regulations themselves.

- Improve regulatory policy to allow regional approaches and priority setting, establish performance-based (rather than technology-based) criteria to include cross-media effects and encourage pollution prevention (for example, modify regulations to make wastewater recycling easier).

- Develop environmental and technology policies that help companies operate in a global economy. Specific suggestions include create environmental policies that lead the world, but not by too much; work to remove international trade restrictions on remanufactured equipment; and eliminate barriers to seeking new technologies internationally (for example on public-private research collaborations that limit participants to domestic firms).

- Invest in municipal infrastructure (nothing was specified but the context was wastewater use and treatment) and materials recycling infrastructure.

- Promote charging customers for the full costs of disposal.

The fact that these recommendations were only suggested by one company could mean a number of things. It could mean that the recommendation is unique to the specific situation of the company or industry sector. Or it could mean that the recommendation is of less importance to industry. This is a limitation of the case-study methodology.

---

[3]During the development of the National Environmental Strategy, *Bridge to a Sustainable Future*, industry's and other stakeholders' (NGOs') policy priorities were gathered (the emphasis during this exercise was much broader than R&D policies). Participants suggested using public-private partnerships in three broad areas. Public-private partnerships with multiple stakeholders (i.e., federal agencies, communities, industries, and regulatory groups) were seen as effective mechanisms to seek solutions that meet environmental and economic objectives, coordinate research priorities, and remove existing barriers to investment, commercialization, and diffusion. Specific cooperative R&D efforts were also suggested, such as one to develop industrial ecology concepts. To make partnerships administratively less costly, industry and other stakeholders recommended expanding the use of cooperative research and development agreements (CRADAs), legislating fewer restrictive antitrust laws, and loosening Federal Advisory Committee Act (FACA) restrictions. Regional alliances to improve technology diffusion and commercialization potential were also recommended (Lachman, Lempert, Anderson, and Resetar, 1995, pp. 33–35).

Chapter Five

CONCLUDING REMARKS AND OBSERVATIONS

WHERE COMPANIES ARE INVESTING IN ENVIRONMENTAL TECHNOLOGY R&D

What have we learned regarding how research-intensive companies are rethinking investments in environmental technologies and where they are likely to invest? In terms of the technology categories, the companies are clearly emphasizing avoidance technologies or those that improve yield, generate desirable by-products, increase energy efficiency, improve the resource efficiencies of products and processes. Investments in control technologies are made only if equipment is not available from other sources. These are straightforward to justify (regulations require it), but investments are minimized. Remediation technologies were important to only one firm, while monitoring and assessment technologies were mentioned, but barely.

Leading companies recognize that while a small portion of their R&D is made exclusively for environmental reasons (estimates ranged from 1 percent to 15 percent), the majority of their R&D investments have an environmental component. They are investing in research and technology to improve the resource efficiency of their products and manufacturing processes because it is cost-effective to do so. R&D investments are made for many business reasons (cost reduction, market access, product improvement, etc.), of which environmental considerations are only one. The case studies have shown however that environmental considerations are being elevated during the R&D planning process because they have linked these concerns to the primary business objectives. For example, Xerox is achieving annual cost avoidance in the hundreds of millions of dollars by reusing equipment; Intel will achieve administrative permitting cost reductions in addition to improving its ability to rapidly change manufacturing processes by investing in technologies to reduce its air emissions; and DuPont can potentially save three to five billion dollars for the next 50 percent improvement in waste per pound of product.[1] All the companies believe there are still many cost-effective improvement opportunities.

In addition, if markets exist, companies will undertake even radical change to satisfy them. All the companies monitor their customers' willingness to pay for environmental attributes. They are tracking global trends in resource scarcity, environmental regulations, voluntary product standards, and customer environmental priorities and needs so they can respond rapidly to emerging markets. Because much of this research deals with competitive knowledge about products and processes, extensive collaboration is less likely than for other technology categories.

Each company is also changing its relationship with suppliers and customers. As described in the definition, environmental technology innovation includes more than technology alone. Organizational and services innovation, such as tighter links between manufacturers, their suppliers, and their customers, may spur additional technology innovations to lessen environmental effects at lower cost or increase technology diffusion. The case-study companies are sharing information regarding environmental regulations, resource scarcity, product standards, and customer environmental priorities with suppliers and customers to develop new products and services with improved environmental features. In the past, these links have been important for expanding the use of recycled materials and product remanufacture as well as for creating alternative, more environmentally benign chemicals. In some cases, companies require suppliers to avoid certain chemicals and to develop processes to manage environmental issues.

In the end, we did not find any systematically collected quantitative data on industrial investments in environmental technology R&D. Our sense is that these case studies illustrate a change in corporate perspectives toward investing in environmental technologies, but not all firms will share this view. The case studies showed that these investments are "large," but only one quantitative estimate of "large" was provided. A few published estimates ranged from 1 percent to 13 percent of R&D is devoted to pollution-control devices and 50 percent of R&D has an EH&S component (Rushton, 1993; Rich, 1993; OTA, 1994). These were calculated in different time periods and were based on different samples. Yet, systematically collected data on environmental technology R&D will improve policymaking. Without information on how much industrial R&D investment has an environmental component, where these investments are

---

[1]The largest element of this estimate is the amount of revenues waste material could have generated if it were product. Waste per pound of product will be reduced with the full range of DuPont's environmental technology investments—yield improvements, sustainable products, reuse/recycle, zero emissions, and co-product development technologies.

being made, and how these investments are changing over time (in response to markets and policies), only a limited understanding of the effectiveness of public policies can be gleaned.

THE CASE STUDIES SUPPORT WHAT WE KNOW ABOUT INNOVATION AND R&D

> "It is the business community that will undertake the principal task of actually delivering solutions."
>
> —*Michael Heseltine, Environmental Secretary, United Kingdom (Schmidheiny, 1992, p. 178).*

Overall industry sees strength in its ability to link inventions to markets and to commercialize new technologies. It relies on a rich science and technology base for future environmental technology innovations. Every company had links to universities to monitor and obtain new expertise. Smaller technology-based companies provided niche capabilities. This suggests that policies that encourage diverse technological approaches by supporting these institutions could lead to radical change and improvement.

In all the cases, one innovation led to another. And in each case, the experience led to additional environmental improvements. In some cases, these were new applications of the technology; in others they were refinements of existing features or practices.[2] For example, DuPont learned to apply its hazardous material handling expertise to operations in other companies. Monsanto learned of the no-till advantages of Roundup® herbicide after the product was used by farmers. And Xerox modified its copy cartridges and customer communications to improve the economics of reuse after experimenting with various recycle/reuse schemes. Since it is impossible to project all conceivable applications or features that become useful to customers, experience is especially important to radically new technologies because part of the innovation process involves creating new markets and experimenting with product features. The current debate over transgenic crops illustrates how experience will inform policy and R&D processes. Institutions or mechanisms to incorporate experience from the marketplace (U.S. and foreign) into the R&D process will improve industrial and federal investment strategies, programs, and projects.

Another observation from the case studies and the literature regarding innovation is that innovation, especially radical innovation, can take a long time—decades or more. Diffusion may take even longer. Given these time frames the

---

[2]RAND has performed extensive research into the benefits of operational experience and technology maturation on weapon systems (Dumond et al., 1994).

benefits of policies to promote new technological approaches might not be fully realized for some time.

FEDERAL POLICIES MUST BALANCE MANY INTERESTS AND FIRM BEHAVIORS

Federal policymakers have a plethora of tools available to promote environmental technologies. Some policies with the most direct connection to R&D investments include federal investment, public-private research partnerships, federal support for high-risk technology demonstrations, use of federal laboratories' capabilities and resources, research and experimentation tax credits, and federal-state collaboration on research efforts. Yet others, such as environmental regulations, product liability laws, green labeling programs, federal and state procurement criteria, foreign aid and technology assistance, education and training investments, and programs to collect and disseminate environmental information, clearly influence environmental technology R&D investments as well.[3]

Ultimately the mix of policies implemented must balance the cost-effectiveness of the various tools, stakeholder interests, and the variety of industrial behaviors toward environmental technologies investments.

Discussion of Select Policies Firms Would Like to See

The companies' policy recommendations themselves show what innovation theory suggests—that multiple policy tools are necessary to stimulate the creation and implementation of new ideas and environmental technologies. While certain policies received more emphasis than others, each company had a series of suggestions that addressed the many components of the innovation process.

Scientific Information. Policies that reduce investment uncertainty by improving and expanding the available information regarding environmental issues will help environmental technologies. If a market is identified, we have seen that corporations can be patient. However, determining the right time to invest involves predictions about markets, technologies, sociopolitical conditions, and regulations (Schneiderman, 1991, p. 56; Myers and Rosenbloom, 1996). Sometimes firms guess incorrectly—either too early or too late—and as a result may

---

[3]This is not an exhaustive list of potential federal policy tools, but it does provide an illustrative overview. For more specifics on federal policies, see NSTC (1994) or OTA (1994).

lose significant market share or go out of business.[4] Unlike other guesses about the future, if guesses about environmental issues are wrong, additional costs stemming from irreparable damage to human health, communities, or natural systems may arise. In all our discussions, firm representatives placed heavy emphasis on the use of science to inform public policy and environmental decisionmaking. There are many dimensions to this statement. Not only does it require investments in the appropriate sciences, but it also requires having institutions capable of communicating knowledge to stakeholders, incorporating this knowledge into regulations, having the necessary decision tools and criteria, etc.

All kinds of information is required to inform decisionmaking. For example, as we have seen from the Intel Project XL agreement, the inability to define superior environmental performance precisely, the lack of complete toxicological data and comprehensive risk-assessment methods, and the inability to monitor actual environmental performance closely have contributed to differing interpretations about the Intel plan and its enforcement. At the firm level, analysis tools encode priorities and make environmental decisionmaking easier. Each of the firms interviewed have been developing a variety of decision and analysis tools, which in turn influence the environmental technology investments that are made. These tools require information; yet a recent EPA review, cited by Vice President Al Gore, found that of the 3,000 chemicals used most widely in the United States, complete health effects data exist for only 7 percent, and well over half had no data at all (White House, 1998). Information and data—on the full cost of materials and energy use or environmental impacts, etc.—will improve investment decisions made by industry and others. Finally, better information on what is an environmentally preferable or sustainable product gives industry and its customers the opportunity to make better choices. We have heard that the lack of established product environmental criteria opens a company that promotes its environmental profile too vigorously to criticism and unfavorable recognition. Because so much of the information needed to guide decisionmaking is beyond the scope of an individual company or an industry and requires open, fully-vetted standards, there is a clear government role here.

The other dimension to incorporating science into environmental decisionmaking requires scientifically sound and well-managed institutions to assess the science, collect data, communicate findings, and balance competing inter-

---

[4]For an example of how Xerox miscalculated its early invention of the computer and other information technologies see Smith and Alexander (1988). Motorola's late introduction of digital cellular phones cost it market share and access to capital. Intel's new challenge to address restructuring of the computer market to emphasize low-cost machines (due in part to the growth of the Internet) is leading to questions about its ability to continue to dominate semiconductor manufacturing.

ests. One company suggested that public and industrial confidence in regulatory processes and agencies will help new technologies and new approaches by ensuring public confidence and acceptance of new technologies.

Intellectual Property Rights and Diffusion. Industry would like to see the federal government work internationally to ensure global enforcement of patent laws to protect intellectual property. However, because environmental technologies have a large public-good aspect, rapid diffusion is desired to realize the environmental benefits from the new technology more quickly. Diffusion may also spur additional innovations, leading to new benefits. While strong intellectual property rights may draw additional R&D investment, they could also slow diffusion by limiting access to a new environmental technology or by raising its price, although widespread licensing can mitigate this problem. As a result, new systems for protecting intellectual property must balance these issues. The economic literature suggests that the current patent system is ineffective because it assumes homogeneity in industrial sectors' structures and mechanisms for technological change. Some have suggested that a new system that takes into account the differences in industrial sectors and particular innovations should replace the current system.[5]

Environmental Regulations Have Played a Role

Environmental regulations have had negative and positive effects on innovation. The literature discusses how environmental regulations can add an additional element of uncertainty to investment decisions and limit long-term innovation because of uncertainty regarding the form of the regulation, its enforcement, the administrative burdens of verifying performance and modifying permits, market segmentation resulting from differing standards, and the liabilities associated with potential performance failure. Some of these aspects of the regulatory process reduce the benefits to technology providers of investing in new technologies, and some are disincentives for regulated companies to be the first adopter of new compliance technology. The aversion to being first adopter of new compliance technology was expressed in our interviews as well. In contrast, as companies link environmental issues to corporate strategy, competitive advantage will encourage some companies to be an early adopter of new technology.

However, environmental regulations have clearly influenced innovation at these firms in a variety of ways. The TRI, Clean Air Act Amendments (CAAA), and other regulations have made avoidance more cost-effective and have increased awareness regarding emissions and the benefits of emissions

---

[5]For more on this topic, see Thurow (1997) and Merges and Nelson (1990).

reductions. The TRI may have stimulated these companies to look at emissions and wastes as opportunities to save money, rather than assuming that emissions cannot be changed. Costly permitting, in terms of time and money, is causing some to rethink their approach to compliance. Emissions controls, hazardous-waste management, and other regulations are stimulating them to rethink their own and their customers' material and energy flows. Where there is rapid product obsolescence, the opportunity cost of compliance with regulations can be greater than the investment required for pollution prevention because of lost market opportunities and permitting processes that are longer than the product life cycle. The findings from a study of Japanese-affiliated manufacturers suggest that innovative companies, which are continuously changing their products and processes, more frequently incorporate new avoidance process technologies because they can more easily absorb the fixed costs associated with these changes (Florida, 1996, p. 99). We have seen that regulations have created, and will continue to create, markets for environmental technologies by changing the cost structure of emissions.

Well-managed and scientifically rigorous environmental regulatory practices are an important policy tool to negotiate the risk and uncertainty associated with new environmental technologies and thereby speed their diffusion. From the companies' point of view, public confidence in the regulatory process is as important as the scientific practices employed and will improve public acceptance of new technologies and new approaches. However, the extent to which regulations and not other management practices, such as total quality management, ISO 14000, or supplier management, influence environmental R&D investments is unclear from the information collected in these interviews. And despite the fact that our observations confirm that regulations do influence companies' behavior, Intel, which is heavily involved in Project XL and other regulatory experiments, was the only one to mention any regulatory reinvention experiments at the federal or state level. Most mentioned in passing that regulations serve good purposes but that sometimes the administration of regulations can be burdensome. Other than mandatory and voluntary product standards, *no one mentioned the future of environmental regulations to encourage innovation*—other than to say that regulations should incorporate good science and allow pollution prevention. In contrast, there was a lot more discussion of sustainability and design-for-environment to stimulate environmental investments and innovation.

A Collection of Policy Tools Will Likely Be Necessary

Our discussions left us with the clear impression that effective federal policies to promote environmentally improved technologies will require multiple policy tools. It cannot simply focus on one, such as environmental regulations. Regulations, of course, play a role. They can be effective means to include externalities into firm cost structures and have stimulated technology innovation, at least in the short run. However, the landscape of environmental technology R&D is complex, and no single tool will sufficiently foster the full range of environmental technology R&D investments. Federal investments in science improve the knowledge base for environmental priority-setting, and stakeholder processes will help create consensus. Support for university-based research may foster dramatically new technological options as well as train the next generation of industry-based researchers and engineers. Raising consumer awareness will increase demand for products that have improved environmental performance and help with environmental priority-setting. Public-private partnerships may help leverage funds to address common technology issues or may be effective means to build consensus among stakeholders.

Each of these policy tools address different elements of the innovation process. Many are mutually reinforcing. We have heard that judicious use of public-private partnerships, broad support of university and national laboratory research, and brokering international agreements to protect intellectual property all can play a role in leveraging the substantial investment industry makes in R&D.

What We Did Not Hear

Several topics are noteworthy simply because we did not hear much about them. In some instances, this may not be surprising since industry *leaders* were interviewed. Tax policy was not mentioned at all—neither pollution taxes nor research and experimentation tax credits. Nothing was said about emissions trading or product liability laws. ISO 14000 was not explicitly linked to innovation although all the firms establish their own environmental performance goals and seek continuous improvement.

While we observed that regulations influenced companies, Intel was the only company to suggest additonal regulatory experiments.[6]   In contrast, much more discussion revolved around sustainability and design-for-environment to stimulate investments and innovation.

---

[6]Two other companies mentioned specific changes to regulations, such as making wastewater recycling easier or incorporating pollution prevention, but these did not receive as much emphasis as other recommendations.

ADDITIONAL POLICY RESEARCH

The recommendations themselves suggest areas for policy research. These include criteria and data for sustainable products; product systems modeling, including material and energy flows; or economic consequences of technology choices integrated with environmental effects. Other related areas might include research on consumer behavior, for example, product choice and decisionmaking, environmental communications, customer environmental awareness, new technology acceptance, and willingness-to-pay studies.

Perhaps most critical is to design and implement a survey to systematically collect data on environmental technology R&D investments by a representative sample of industry over time. These data would provide the critical information required to shape cost-effective federal policies.

Since companies from three industrial sectors were interviewed, additional case studies could continue to provide this vital link to industry views in a structured and consistent manner. Carefully selected cases can enhance federal decisionmakers' understanding of industry's interests in federal policy. Information from other industrial sectors, such as consumer products, services, transportation, aerospace, petroleum refining, or energy production, could provide a more complete picture of the breadth of industrial priorities. Cases within the original set of industrial sectors of chemicals, electronics, and biotechnology would give policymakers a better feel for the variety of firm behaviors within a sector and may help to clarify which activities are sector- or firm-specific. Closer attention to the sources of innovation would improve our understanding of the relationship between these large problem-holders and the smaller companies that can provide solutions as well.

Another area that we heard a lot about is "good science." Because this is such a ubiquitous topic, further study on what constitutes "good science" and how to ensure that it is incorporated into industrial and public environmental decisionmaking and policy development seems warranted. A critical dimension is how scientific information is communicated to all stakeholders. An interdisciplinary study—for example a retrospective of bovine growth hormone or a prospective study on future of transgenic crops—that addresses how science was used, what aspects were controversial, where gaps were, how the stakeholders interpreted it, would inform future environmental science and technology policy.

Each of the themes in Chapter Four could be used to develop specific policy packages either through small, focused workshops (face-to-face or worldwide-web based). These policy packages could be evaluated against various stakeholder priorities using gaming techniques (gaming allows policymakers to test policies by simulating the actions of various actors). Since industry is developing measures of eco-efficiency before the concept of sustainability has been fully defined, some key questions include the following:

- How far will resource efficiency take us toward sustainability?

- Are the eco-efficiency measures of the WBCSD and others the right measures to manage for sustainability and will they instill the right kind of change?

- What do they tell us about the R&D investments companies are likely to make?

To sustain economic and population growth while balancing human and environmental health, technology innovation may help achieve the orders-of-magnitude improvement needed. Sustainability is clearly an idea that industry leaders are wrestling with. Jack Krol (1997), former DuPont CEO outlined three challenges for the chemical industry. These challenges are value creation, technology, and sustainability. These challenges transcend the chemical industry—they are challenges for *all* industry, *all* government, and *every* citizen.

ENVIRONMENTAL TECHNOLOGIES DEFINITION

The following definitions have been taken from *Technology for a Sustainable Future* (NSTC, 1994, p. 9, and pp. 42–43).

Environmental Technology. An environmental technology is a technology that advances sustainable development by reducing risk [of harm to human health or the environment], enhancing cost-effectiveness [of achieving environmental protection], improving process efficiency, and creating products and processes that are environmentally beneficial or benign. The word "technology" is intended to include hardware, software, systems, and services.[1]

The categories of environmental technologies are as follows:

- **Avoidance.** Avoidance technologies avoid the production of environmentally hazardous substances or alter human activities in ways that minimize damage to the environment.

- **Monitoring and assessment.** Monitoring and assessment technologies are used to establish and monitor the condition of the environment, including releases of pollutants and other natural or anthropogenic materials of a harmful nature.

- **Control.** Control technologies render hazardous substances harmless before they enter the environment.

- **Remediation and restoration.** Remediation technologies render harmful or hazardous substances harmless after they enter the environment. Restoration technologies embody methods designed to improve ecosystems that have declined because of naturally induced or anthropogenic effects.

---

[1]Clarifications in brackets added.

DATA ON ENVIRONMENTAL AND R&D ACTIVITIES

The data in Tables B.1 and B.2 present some of the information collected in order to identify candidate companies for the case studies. The rows are various industrial sectors and the columns contain data on environmental and R&D expenditures, activities, and reputations.[1] Table B.1 presents available environmentally related information including sector pollution abatement expenditures, sector TRI emissions and the names of the top ten emitting companies, participation in management initiatives and voluntary programs, reputation for leader or laggard, and award-winners. Pollution abatement expenditures for capital equipment and operations were collected by the Department of Commerce until 1994. These are shown in the third column of the table. TRI emissions reports, required by the Emergency Planning and Community Right-to-Know Act of 1986, contain information on emissions to air, land, water, and off-site transfers for more than 600 chemicals. The fourth column shows these emissions for the entire industrial sector as well as the top 10 emitters overall. Column five contains information on environmental reputation. Some information was reported in two issues of *Fortune* magazine—as part of its annual survey of corporate reputations in a variety of areas and in a special issue. The special issue established reputation for a select number of companies based on a more detailed analysis of 20 criteria that included TRI releases adjusted for sales, TRI emissions reductions, comprehensiveness of environmental report, environmental awards, employee incentives, serious violations, and ratings by leading environmental groups. Other criteria, which received less weight, were number of Potentially Responsible Party (PRP) sites, reuse and recycling of hazardous and solid waste, participation in voluntary programs. Column six contains companies involved in GEMI, which is an industry organization that seeks to develop information and tools to improve environmental management. Column seven shows winners of the McGraw-Hill EPA Environmental Champion Award. Winners were selected based on their

---

[1]In an effort to simplify the tables, the industrial sectors, or rows, for which data were not readily available were deleted.

success in the 33/50 program, which is a voluntary pollution-prevention initiative that targets 17 toxic chemicals for reductions. The goals were an interim goal of 33 percent reduction in 1992, and an ultimate goal of 50 percent reduction in 1995, as measured against 1988 Toxics Release Inventory data. Column eight shows companies active at the time in various EPA voluntary programs—Project XL, Environmental Leadership, and WAVE programs. Column 9 shows companies rated as environmental laggards in *Fortune*.

Table B.2 presents information on sector R&D intensity and companies with the largest R&D expenditures, reputations for quality or innovativeness, and winners of the total quality management award—the Malcolm Baldrige Award. R&D percent of sales is taken from an NSF survey of industry. Column three also lists the top five spenders on R&D in each industrial sector and the amounts they spent in 1994. Column four presents information on reputation as a leader in R&D based on two studies. The first was a survey of firms with R&D expenditures greater than $100 million and the top 10 companies mentioned by peers were identified. The second reported on 22 firms with the most nominations with positive perceptions of their R&D strategic planning processes. A total of 270 firms were nominated. The survey was sent to R&D managers and nominations were from organizations inside and outside the industrial sector. Column five presents information on reputation as an innovator based on studies of companies presented in the literature as noted at the bottom of the table. The last column presents the Malcolm Baldrige Award winners. Managed by the National Institutes of Standards, this award was established by Congress in 1987 to promote awareness about quality management, to recognize quality achievements of U.S. companies and to publicize achievements. The award criteria are: (1) leadership; (2) information and analysis; (3) strategic planning; (4) human resource development and management; (5) process management; (6) business results; and (7) customer focus and satisfaction. Heavy emphasis is placed on business achievement as demonstrated through quantitative data.

Table B.1

Industrial Sector and Firm Environmental Data

Industrial Sector	SIC	Pollution Abatement Expenditures (equipment & operations)[a]	Sector Rank in 1994 TRI Emissions and Top 10 Emitters[b]	Environmental Reputation as a Leader[c,d]	Global Environmental Management Initiative Member[e]	33/50 Program Award Winners[f]	Voluntary Program Participants: EPA's Project XL,[g] Environmental Leadership program,[h] and WAVE[i]	Environmental Reputation: Laggard[b,c]
MANUFACTURING								
Construction	15, 16, 17							Morrison Knudson[c]
Food, kindred, and tobacco products	20, 21	1,863	11		Anheuser-Busch, Coca-Cola, Coors,	Sandoz[f]	Anheuser-Busch[g]	
Paper and allied products	26	2,561	6 International Paper, Georgia-Pacific				Weyerhauser,[g] Simpson Tacoma Kraft Company[h]	International Paper,[c] Lousiana Pacific[c]
Industrial chemicals	282, 284-289	7,138 (includes pharmaceuticals)	1 DuPont, Monsanto	Clorox,[c] Dow,[c] HP Fuller,[c] 3M[d]	Dow, DuPont, Eastman Kodak, Olin, PPG Industries, Union Carbide	3M, Aristech Chemical, DuPont, Eastman Kodak, Henkel, Milliken, Rohm and Haas, Sterling Chemicals	3M,[g] Ciba-Geigy,[h] OSi[g], Union Carbide[g]	DuPont,[b] Monsanto[b]
Pharmaceuticals	283	7,138 (includes industrial chemicals)	1		Bristol-Myers Squibb, Merck,	Merck, Sandoz, SmithKline Beecham, Upjohn	Merck[g]	American Cyanamid[b]

Table B.1—continued

Industrial Sector	SIC	Pollution Abatement Expenditures (equipment & operations)[a]	Sector Rank in 1994 TRI Emissions and Top 10 Emitters[b]	Environmental Reputation as a Leader[c, d]	Global Environmental Management Initiative Member[e]	33/50 Program Award Winners[f]	Voluntary Program Participants: EPA's Project XL,[g] Environmental Leadership program,[h] and WAVE[i]	Environmental Reputation: Laggard[b, c]
Petroleum refining and extraction	13, 29	5,707	7		Amoco, Occidental, Tenneco	Amoco, Halliburton Energy Services		BP America[b]
Primary metals	33	2,808	2			Bethlehem Steel, Reynolds Metals, U.S. Steel Group of USX		USX[c]
Electronic and other electrical equipment	36 (35 computers)	1,135	3	AT&T,[c] Apple,[c] Digital Equipment,[c] IBM,[c] Xerox[c]	Apple Computer, AT&T, Digital Equipment Corp	Lockheed Martin	HADCO,[g] IBM,[g] Intel Corp.,[g] Lucent Technologies (formerly AT&T Microelectronics),[g] Molex Inc.,[g] Motorola[h]	GE[c]
Auto manufacturers	371	1,816	5 General Motors		Allied Signal			
Aerospace	376		5		Allied Signal, Boeing	Lockheed Martin		Boeing[c]
Other manufacturing	22, 26, 24, 25, 27, 30, 31, 32, 34			Herman Miller[c] (furniture), Levi Strauss[d]		Milliken (textiles), PPG Industries (glass)		Maxxam[c] (metals), USX[c] (Steel)
Retail trade/ consumer products, etc.	50			J&J[d]	Colgate-Palmolive, J&J		Gillette Company[h]	Kmart[c]

Table B.1—continued

Industrial Sector	SIC	Pollution Abatement Expenditures (equipment & operations)[a]	Sector Rank in 1994 TRI Emissions and Top 10 Emitters[b]	Environmental Reputation as a Leader[c,d]	Global Environmental Management Initiative Member[e]	33/50 Program Award Winners[f]	Voluntary Program Participants: EPA's Project XL,[g] Environmental Leadership program,[h] and WAVE[i]	Environmental Reputation: Laggard[b,c]
SERVICES								
Engineering, accounting, research, management, and related services	87					Halliburton Energy Services, Laidlaw Environmental Services	WMX Technology[h]	
Tourism	70, 78, 79, 84,				Anheuser-Busch		Hyatt Hotels,[i] ITT Sheraton,[i] Saunder Hotel Grp,[i] Westin Hotels and Resorts[i]	

[a] Figures are expressed in FY 1994 millions of dollars. DoC, *Pollution Abatement Costs and Expenditures: 1994.*

[b] http://www.epa.gov accessed September 1996.

[c] *Fortune,* "Who Scores Best on the Environment," July 26, 1993. Corporations listed as most improved include: Ciba-Geigy, HP, J&J, SC Johnson and Son, 3M, Nalco Chemical, Polaroid, Shell Oil, Sun, Union Camp.

[d] *Fortune,* "Corporate Reputations," March 6, 1996. General reputations based on weighted value of several categories. Environmental combined with community responsibility.

[e] Global Environmental Management Initiative membership, 1994. Companies not included in tables are Browning Ferris, Consolidated Rail Corp., Duke Power Company, Florida Power and Light Co., The Southern Company, WMX Technologies, Inc.

[f] McGraw-Hill and EPA Environmental Champion Awards for 1995 (first annual). U.S. Department of Energy was also an award winner.

[g] Firms accepted for development of a final project agreement as of June 26, 1996 within EPA's voluntary Project XL program. Includes industry submissions only. Berry Corporation not included in table.

[h] Environmental Leadership program, http://es.inel.gov/elp/ accessed September 1995. These companies are involved in the pilot phase. John Roberts Printing, Salt River Project, McClellan Air Force Base, Puget Sound Naval Shipyard, Arizona Public Service, Ocean State Power, and Duke Power are also participating.

[i] http://es.inel.gov/partners/wave/wavesum.html accessed September 1995. The charter members of WAVE (Water Alliance for Voluntary Efficiency), which is designed to focus attention on the value of water and the need for its efficient use. Voluntary partnership agreements between EPA and the logging industry seek to save money by reducing water pollution and use by 15–30% or more. WAVE also encourages hotels and motels to install water-efficient equipment wherever profitable and practical.

Table B.2

R&D: Industrial Sector and Firm Data

Industrial Sector	SIC	1994 R&D % of Sales[a] and Top 5 Companies within the Sector with the largest R&D expenditures[b]	Reputation as R&D Leader [c,d]	Reputation as an Innovator[e,f,g]	Malcolm Baldrige Award Winners[h]
MANUFACTURING		3.6			
Construction	15, 16, 17	NA PPG Industries 218.1 Owens-Corning fiberglass 64.0 American Standard 39.0 Tecumseh Products 27.8 Valspar 27.0		Enron[c] (pipelines)	Armstrong World Industries Building Products (1995), Granite Rock Co. (1992),
Food, kindred, and tobacco products	20, 21	0.5 Pioneer Hibred Int'l. 113.7 Campbell Soup 78.0 Kellogg 71.7 Ralston Purina Grp 69.3 General Mills 63.6		Phillip Morris[f]	
Paper and allied products	26	0.5 Kimberly-Clark 167.1 International Paper 102.6 Union Camp 49.2 James River Corp. 47.0 Weyerhauser 47.0 Manville 39.1			
Industrial chemicals	282, 284-289	4.6 Dow 1,261.0 DuPont 1,047.0 Monsanto 609.0 Rohm and Haas 201.0 Eastman Chemical 167.0 Others 3M 1,054.0 (general mfg)	Dow,[d] Dupont,[c, d] 3M[c, d]	3M,[f] Monsanto[e]	Eastman Chemical Company (1993), Milliken & Company (1989) Wallace Co., Inc. (1990), (pipes, valves, fittings)

Table B.2—continued

Industrial Sector	SIC	1994 R&D % of Sales[a] and Top 5 Companies within the Sector with the largest R&D expenditures[b]	Reputation as R&D Leader[c,d]	Reputation as an Innovator [e,f,g]	Malcolm Baldrige Award Winners[h]
Pharmaceuticals	283	10.2 Merck 1,230.6 Pfizer 1,139.4 Bristol-Myers Squibb 1,108.0 Lilly (Eli) 897.1 American Home Products 817.1 Others (med prod & serv) J&J 1,278.0 Abbott Labs 963.5	Merck,[c, d] Glaxo-Wellcome[c]	Merck,[f] J&J[f]	
Petroleum refining and extraction	13 , 29	0.8 Exxon 558 Schlumberger 418.9 Mobil 275.0 Amoco 225.0 Chevron 179.0	Amoco,[d] Mobil,[d] Shell[c, d]	Enron[c] (pipelines)	Wallace Co., Inc. (1990), (pipes, valves, fittings)
Primary metals	33	0.6 Alcoa 125.8 Reynolds Metal 38.0 Bethlehem Steel 24.0 USX 23.0 Kaiser Alum. 16.7 Maxxam 16.7			Globe Metallurgical Inc. (1988),
Electronic and other electrical equipment	36 (35 computers)	5.9 Intel 1,111.0 Hughes Electronics 699.3 Texas Instruments 689.0 Honeywell 319.0 Emerson Electric 298.2	Apple,[d] IBM,[c, d] GE,[d] Hewlett-Packard,[c, d] Intel,[d] Motorola,[d] Sony[c, d]	Digital Equipment Corp.,[e] General Electric,[f, g] Hewlett-Packard(5 & 10), IBM,[f] Intel,[c, e] Motorola,[f, g] Sony[f]	GTE Directories Corp (1994), IBM Rochester (1990), Marlow Industries (1991) (electronics equip/heat exchangers), Motorola (1988), Solectron Corp. (1991), Xerox- Bus Prod & Sys (1989) (electronics equip/PCB assembly)

Table B.2—continued

Industrial Sector	SIC	1994 R&D % of Sales[a] and Top 5 Companies within the Sector with the largest R&D expenditures[b]	Reputation as R&D Leader[c,d]	Reputation as an Innovator[e,f,g]	Malcolm Baldrige Award Winners[h]
Auto manufacturers	371	5.9 (incl aerospace) GM (incl Hughes) 7,035.8 Ford 5,214.0 Chrysler 1,300.0 Allied Signal 318.0 Goodyear Tire 3,41.3		Ford[f]	Cadillac Motor Car Co. (1990), Wainwright Industries, Inc. (1994), (parts supplier)
Aerospace	376	5.9 (incl auto manufacturers)	Boeing[d]	Boeing[f]	TI Inc.-Defense Systems & Electronics Group (1992)
Other manufacturing	22, 26, 24, 25, 27, 30, 31, 32, 34	AT&T 3110.1 (telecomm) Xerox 895.0 (office equip)	AT&T,[c, d] Xerox[d]	Corning,[e, g] Xerox[e]	Ames Rubber Corporation (1993), AT&T Consumer Communications (1994)/Network Systems Grp (1992)/Universal Card Serv (1992), Corning Telecomm Prod Divi (1995)
Retail trade/ consumer products, etc.	50	Consumer Products: P&G 1.059.0 Eastman Kodak 859.0 Philip Morris 435.0 Pepsico 152.0 Whirlpool 152.0 Gillette 136.9	P&G[c, d]	Rubbermaid,[c] P&G,[f] Philip Morris,[f] Wal-Mart,[f] Nordstrom[f]	

Table B.2—continued

Industrial Sector	SIC	1994 R&D % of Sales[a] and Top 5 Companies within the Sector with the largest R&D expenditures[b]	Reputation as R&D Leader[c,d]	Reputation as an Innovator[e,f,g]	Malcolm Baldridge Award Winners[h]
SERVICES					
Engineering, accounting, research, management, and related services	87		Microsoft[d]		
Tourism	70, 78, 79, 84,			Marriott,[f] Walt Disney[f]	Ritz-Carlton Hotel Co. (1992)
Financial	60-67			American Express,[f] Citicorp[f]	

[a]NSF, *Survey of R&D in Industry: 1994*, Table A-17. Paper and pulp value is 1991, Chemical value is 1992.

[b]Figures are expressed in FY 1994 millions of dollars. *Business Week*, "1995 R&D Scoreboard," July 3, 1995.

[c]Roberts, Edward (MIT), "Benchmarking the Strategic Management of Technology—II" *Research Technology Management*, March-April 1995, pp. 18-26. Top 10 firms were reported.

[d]Michael Menke, Strategic Decisions Group, Menlo Park, California. Three other organizations identified but not included in the table are Pacific Gas and Electric, EPRI, and Southern California Edison.

[e]*Wellsprings of Knowledge*, Dorothy Leonard-Barton, Harvard Business School Press, 1995.

[f]*Built to Last: Successful Habits of Visionary Companies*, James C. Collins and Jerry I. Porras, HarperCollins, 1994. The authors describe visionary companies as "premier institutions—the crown jewels—in their industries, widely admired by their peers and having a long track record of making a significant impact on the world around them." The criteria they used for choosing visionary companies were: (1) premier institution in its industry; (2) widely admired by knowledgeable business people; (3) made an indelible imprint on the world in which we live; (4) had multiple generations of chief executives; (5) been through multiple product (or service) life cycles; (6) founded before 1950.

[g]*Winning in High-Tech Markets: The Role of General Management*, Joseph Morone, Harvard Business School Press, 1993.

[h]*Malcolm Baldrige National Quality Award: Profile of Winners*. Other winners include: Federal Express Corporation (1990), Westinghouse Electric Corporation—Commerical Nuclear Fuel Division (1991), and Zytec Corporation (1991).

DUPONT: BETTER THINGS FOR BETTER LIVING

DUPONT IS A LARGE MULTINATIONAL COMPANY

DuPont is a large, multinational company with products in eleven broad categories—agriculture (pesticides and herbicides), biotechnology, chemicals, electronic materials, energy and petroleum,[1] fibers (apparel and general), films and resins, medical products, printing and publishing, safety and environmental management services, and quality management and technology services (DuPont, 1995a). DuPont's products are used in a variety of industries including aerospace, agriculture, automotive, construction, electronics, packaging, refining, and transportation markets. Some of these areas have more rapid rates of product innovation than others. For example, 30 percent of the 1995 revenues in electronic materials were from products developed within the previous three years. Twenty percent of the films and resins' revenues were from products developed within the previous three years.[2] Other areas, such as nylon and titanium dioxide, are more stable. For these products, DuPont competes on production efficiencies and customer applications. R&D may be geared toward gaining knowledge for new applications.

DuPont's 1997 revenues were $45.1 billion, up from $43.8 billion in 1996. Fifty percent of these sales were from outside the United States. There were 105,000 employees worldwide, 30 percent of whom live and work outside the United States. In 1994, a total of 175 manufacturing plants (150 chemical plants and 25 petroleum refining and natural gas) and 80 R&D facilities and service centers were located in 70 countries (see Table C.1) (DuPont, 1996a and 1998a). DuPont is in the process of selling its oil and gas business (Conoco), which accounted for almost half of its revenues in 1997.

---

[1]On May 11, 1998 DuPont announced that it plans to divest its Conoco energy subsidiary (DuPont, 1998b).

[2]DuPont, http://www.dupont.com.

Table C.1

DuPont Revenues and Facilities (dollars in billions)

Business Area	1994	1995	1996	1997
Chemicals				
Chemicals	3.8	4.2	4.1	4.3
Fibers	6.8	7.2	7.2	7.7
Polymers	6.3	7.0	6.7	6.8
Petroleum Refining and Natural Gas	16.8	17.7	20.2	21.0
Diversified and Life Sciences Businesses	5.6	6.0	5.6	5.3
Total Revenue	39.3	42.1	43.8	45.1
R&D Investment	(1.2)	(1.1)	(1.0)	(1.1)
R&D Intensity	3.0	2.6	2.2	2.4[a]

[a]DuPont's R&D intensity would rise to an estimated 3.9 percent of revenues without the Conoco oil and gas division (assuming its R&D investments follow the petroleum industry's average of 0.8 percent of revenues).

SOURCES: DuPont Corporation, DuPont Safety, Health and the Environment: 1995 Progress Report, and DuPont 1996 Annual Report to Shareholders, 1997 Annual Report to Shareholders.

Largely because of its TRI emissions DuPont was identified in a *Fortune* study as an environmental laggard (Rice, 1993).[3] In 1994 DuPont was the largest emitter of Toxic Release Inventory (TRI) chemicals in the country, legally emitting over 203 million pounds of waste to the air, soil, and water.[4] It was noted by DuPont that 80 percent, or 166 million pounds, of these emissions were injected into deep wells, a method approved by the EPA. In addition, 112 million pounds of these emissions were hydrochloric acid, which was de-listed from the TRI in 1995. The authors wish to note that TRI emissions are reported as pounds of emissions and are not normalized for hazard, production volume, or line of business. However, DuPont won a McGraw-Hill award for the com-

[3]The 20 different categories include TRI releases adjusted for sales, TRI emissions reductions, comprehensiveness of environmental report, environmental awards, employee incentives, serious violations, and ratings by leading environmental groups. Other categories received less weight, including number of PRP sites, reuse and recycling of hazardous and solid waste, and participation in voluntary programs. *Fortune* analysts describe the DuPont environmental report as "first-rate" and suggest that the "well-coordinated pollution-prevention programs are making inroads." However, because DuPont was the largest emitter of TRI chemicals and is a potentially responsible party at more than 100 Superfund sites, the company was classified as laggard.

[4]Note: waste emitted is measured in pounds, which is a very incomplete measure of environmental impact because weight does not measure overall toxicity. Moreover, many chemical plant emissions are dilute wastewater, which weighs a lot but has lower toxicity. Recognizing this, some companies have been able to reduce emissions simply by reducing the water content of these waste streams. The next largest emitter of TRI chemicals in 1994 was ASARCO Inc., which emitted more than 69 million pounds, roughly one-third of DuPont's emissions. (EPA, 1994, Table 1-21, "Top 10 Parent Companies with the Largest Total Releases.")

pany's voluntary manufacturing emissions reductions for 17 chemicals on the TRI list as part of the EPA's 33/50 Program (McGraw-Hill and EPA, 1995).[5]

In terms of environmental activities, DuPont subscribes to both the chemical industry's code of management practices called Responsible Care® and the petroleum and refining industry's program called STEP. The company is an active member and founding member of both the WBCSD, a group of business leaders who seek to integrate environmental protection and economic development, and the Wildlife Habitat Enhancement Council, an organization that helps companies manage underdeveloped and buffer properties for wildlife. DuPont's land legacy program was established in 1994 and targets surplus company properties for habitat preservation. DuPont is also a member of an industry group GEMI.

THE FOCUS OF R&D IS CYCLICAL AT DUPONT

DuPont R&D investments have been fairly stable over the past four years at more than $1 billion per year. This is approximately 2.5 to 3.0 percent of total revenues, although there is considerable variability across the business areas.[6] In any given year stable product lines, such as industrial chemicals, acids, bases, and solvents, might invest in R&D on the order of 0.5 percent of revenues, while new technology products, such as agricultural pesticides, performance polymers, and electronic materials, might invest 10 to 14 percent of revenues (Carberry, 1997). The corporate research center, named the Experimental Station, seeks fundamental research and process-oriented improvements. The business areas on the other hand, control a much larger portion of R&D (compared to corporate control) (Carberry, 1997).

Technology innovations in both products and processes have been instrumental to the growth of the chemical industry. For chemical companies in general, basic research has been more important to this industry sector than other industry sectors (Rosenberg, 1994, p. 191).[7] Between 1930 and the early 1980s there were 63 major product innovations in the chemical industry. Forty of these occurred in the 1930s and 1940s; 20 in the 1950s and 1960s; and 3 in the 1970s to early 1980s. A similar slow-down in process innovations was noted as

---

[5]Winners were selected based on their success in the 33/50 program, which is a voluntary pollution-prevention initiative that targets 17 toxic chemicals for reductions. The goals were an interim goal of 33 percent reduction in 1992, and an ultimate goal of 50 percent reduction in 1995, as measured against 1988 Toxics Release Inventory data.

[6]This value is likely to be closer to 3.9 percent if the Conoco Division (oil and gas) is not included and assuming that division invests at the petroleum industry average of 0.8 percent of revenues.

[7]The data supporting this statement covered the years 1965 to 1984.

well.[8] However, concern about investments in basic research at DuPont have persisted almost since the company was founded.

Charles Stine, DuPont's head of research in 1926, put it this way:

> The sort of work we refer to is work undertaken with the objective of establishing or discovering new scientific facts. . . . The volume of fundamental work is rapidly losing ground as compared with the volume of applied research. In other words, applied research is facing a shortage of its principal raw materials. (Miller, 1997, p. 6.)

It has been more than 70 years since this statement was made, and over these years DuPont R&D personnel and other have reported that the major thrust of the company, with implications for R&D, has followed a fairly predictable cyclic pattern (Miller, 1997; Sharp, 1994). Every 17 to 20 years the corporate R&D management emphasis (for the mix of product lines) fluctuates between focusing on new product discoveries and business expansion and focusing on consolidation and cost-reduction research. DuPont executives anticipated the beginning of another growth or new discovery period beginning around 1996 (Miller, 1997, pp. 26–28).[9] Figure C.1 describes the cyclical pattern to DuPont management emphasis.

Discovery research was estimated to approach 18 percent of the more than $1 billion R&D budget in 1995 (DuPont, 1996b).[10] Despite the prediction that new product-oriented R&D may be the focus for R&D in the next several years, DuPont also expected to invest in R&D to improve the productivity of existing assets, boosting asset productivity by at least 20 percent in the 1995–2000 period (DuPont, 1996b). These investments are equally important to the success of the company—"success in the commercialization of chemical-processing innovations has depended critically upon the productivity gains realized through an improvement process that takes place after an innovation is introduced into the market." (Rosenberg, 1994, p. 198.)

---

[8]Major product advances during these periods include fiberglass, herbicides, flame retardants, epoxy adhesives, high-nitrogen fertilizers, and catalytic converters (Dertouzos, 1989, p. 192).

[9]Note: a typical product life cycle follows a similar pattern. The product life cycle refers to four stages relating product lifetime to sales. In the introduction and growth stages of a product life cycle, sales increase over time. A product reaches maturity when, among other things, sales peak. In the final stage, the decline, sales decrease. A company's R&D will emphasize different factors in each stage of the product's life cycle. In the introduction stage, R&D aims to improve product characteristics, particularly those emerging in the dominant design. In the growth stage nearly to the end of the maturity stage, R&D funding increases to improve product features and gain market share. Process-oriented research becomes more important in this stage as well. In the decline stage, R&D funding decreases as well. In some cases, a small amount of R&D for product modifications may be spent if this will stimulate the market. (Link and Bauer, 1989, pp. 10–12.)

[10]Quoting Joseph Miller, chief technology officer and senior vice president for R&D.

RAND *MR1068-C.1*

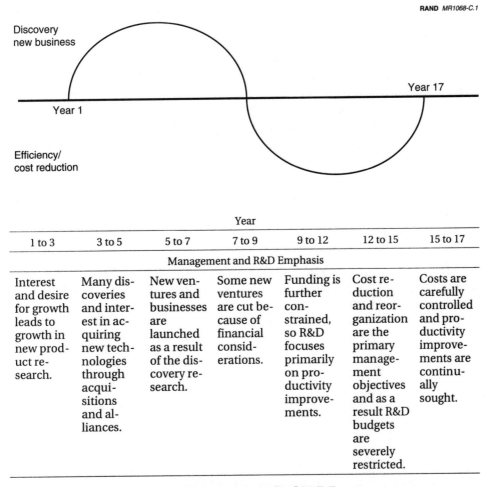

			Year			
1 to 3	3 to 5	5 to 7	7 to 9	9 to 12	12 to 15	15 to 17
			Management and R&D Emphasis			
Interest and desire for growth leads to growth in new product research.	Many discoveries and interest in acquiring new technologies through acquisitions and alliances.	New ventures and businesses are launched as a result of the discovery research.	Some new ventures are cut because of financial considerations.	Funding is further constrained, so R&D focuses primarily on productivity improvements.	Cost reduction and reorganization are the primary management objectives and as a result R&D budgets are severely restricted.	Costs are carefully controlled and productivity improvements are continually sought.

Figure C.1—Dupont's Cyclical R&D Focus

Product and cycle-time excellence (PACE) is the name for DuPont's strategy to maximize outcomes from R&D investments. The PACE strategy emphasizes assessment of the substance of R&D, not improvement of project management per se. Basically, decisions regarding the commercialization potential of the development are made much earlier in the R&D process, before large sums of money and time have been committed. This means that throughout the development process, gains in scientific and technical knowledge are constantly integrated with market information. As a result of the PACE strategy, technologies that may only exist in the laboratory are assessed against a larger number of traditional R&D outcome measures and financial analyses, such as market position, net present value, after-tax income, capital efficiency, product quality, and environmental excellence. Now if a new technology is successful in the critical pilot plant phase, it is twice as likely to be implemented as in the past.

DuPont also actively develops linkages to external sources of R&D. Although these activities represent a small portion of all DuPont's R&D investments—in 1995 it totaled $45 million, or about 4 percent of all R&D—they are growing. The 1995 figure was double the amount spent in 1993. These agreements have primarily been with universities (they have agreements with more than 200 universities). DuPont also has 30 cooperative R&D agreements (CRADAs) with national labs. External sources of R&D are sought to leverage internal R&D by two to ten times, in other words, DuPont would have to spend two to ten dollars internally for every dollar spent on outside sources of technology (DuPont, 1995a).[11] The recent creation of a technology transfer center illustrates the growing importance of this activity.

ENVIRONMENTAL R&D IS BECOMING A STRATEGIC ISSUE, AND SOLVING CUSTOMER ENVIRONMENTAL PROBLEMS IS BECOMING A NEW COMPETENCY

> The challenge facing industry in the 21st century will be to employ sustainable manufacturing practices that reduce waste and emissions while providing competitive advantage and business growth. . . . Incremental changes and numerical goals alone won't get the job done.
>
> —*John Krol, former CEO of DuPont (DuPont, 1996c.)*

Since around 1988, the corporate approach to environmental R&D investments has been changing at DuPont. In the past, environmental R&D was viewed by management as a cost—as serious lost opportunities—and was treated as such. The objectives for environmental R&D were simply to develop or purchase technology at minimal cost to comply with the law. Since compliance was the only goal, environmental technology investments were viewed as a defensive cost maneuver, and, because failure of a new technology could only mean trouble, there were no real incentives to be the first adopter of a new technology. The opportunity costs of expenditures in compliance activities were large—the annual expenditures of around $1.2 billion made on all environmental activities (remediation, capital equipment, training, compliance, etc.) were comparable to R&D investments. But these expenditures would not lead to future markets and revenue streams. Today, however, environmental issues

---

[11]General Motors seeks to leverage internal research with external research as well. They prefer arrangements with a ten to one payback (France, 1997).

are not strictly limited to concern about DuPont's manufacturing emissions. Since 1988, DuPont gradually began to realize that environmental R&D investments could be used to meet strategic business objectives beyond cost containment. These investments, primarily on the products and services side, can lead to competitive advantage because DuPont seeks to help customers solve their problems and meet their own regulatory requirements.

Now seeking new "environmental" markets is a strategic issue for R&D. These areas have evolved as DuPont has extended its environmental expertise and competencies to applications other than its own. In addition to compliance with the law, three new classes of strategically relevant environmental activities draw environmental R&D investments. The first class of activities is driven by DuPont's desire to be the supplier of choice because of a positive environmental image in the marketplace. For example, herbicide products with very low mammalian toxicology are attractive to a large and growing market. DuPont's sulfonylureas class of chemicals,[12] which have been developed from the early 1980s to the present, are examples of environmentally preferable products. An example of what DuPont calls a sustainable process (still in development) is the 3G process for making polyester intermediates. It uses glucose from cornstarch, a renewable, as a feedstock. Its fermentation process requires no heavy metals, petroleum, or toxic chemicals. The microbial by-product can be used as animal feed. And all the liquid effluent is biodegradable. Finally, the 3G polymer created can be regenerated using methanolysis, which is a DuPont proprietary chemical process that "unzips" the molecules. Recombinant DNA technology was used to create a microorganism with all of the enzymes required to turn the glucose into 3G. DuPont anticipates that this process will be cheaper than conventional chemical processes. This technology was developed through an alliance between DuPont and Genecor International (Krol, 1997; Holliday, 1998).

The second class of strategically relevant environmental activities deals with markets where customers require certain environmentally related criteria. These criteria may be recycled material content or disassembly for recycle, etc. For example, DuPont invested in R&D between 1992 and 1993 to determine how to satisfy Federal Express's requirement that envelopes made of Tyvek®, a material stronger than paper, contain 25 percent postconsumer waste. DuPont now recycles the polyethylene in milk and water jugs to make Tyvek® envelopes for Federal Express.

---

[12]Product names include Glean®, Harmony®, Ally®, Express®, and Granstar® (grain), Classic® (soybeans), Londax® (rice), and Accent® (corn). According to DuPont's Director of Environmental Technology R&D, these products offer very low mammalian toxicology. Dose rates have been reduced by a factor of 100 from previously available products.

The third class, which is emerging, is the ability to solve the customer's environmental problems for them. This competency is really the opportunistic application of DuPont's experience in handling toxic and hazardous materials to meet customer needs. For example, DuPont has more than 20 years of experience handling sulfuric acid, a hazardous material. After realizing that it could apply this expertise to its customers' operations, DuPont began studying alternatives to reduce its customers' handling issues associated with such toxic or hazardous materials as hydrogen cyanide. DuPont could do this in a number of ways. The toxic material could be sold and delivered in just-in-time quantities to reduce the amount in the customer's inventory. Or, DuPont could take it one step further and manufacture toxic material at the customer's site to reduce both inventory and transportation risk. Or, DuPont could actually manufacture the customer's component that uses a toxic material and then ship the benign component to the customer. This last option concentrates material handling at a centralized "center of excellence" and takes advantage of DuPont's experience with high-security, low-inventory, and just-in-time manufacturing. Another example is a product DuPont developed for farmers. DuPont developed a water-soluble plastic box liner for several of its products so that farmers could more easily dispense chemicals used for processing. With the new system, the water-soluble packaging is easily removed from the cardboard container and placed into the farmer's equipment, eliminating the need to treat the entire cardboard container as hazardous waste. Thus, helping customers meet their own environmental goals and compliance needs is a new core competency developing at DuPont.[13] In these cases, there is a significant advantage to being the first to meet customer needs.

As a result of this new perspective on environmental investments roughly 95 percent of all R&D investments can be linked to environmental issues in some fashion, either exclusively (15 percent), largely (50 percent), or modestly (30 percent).[14] This change has been institutionalized in three ways as described below.

---

[13]Other examples include DuPont's pollution-prevention actions, which have led to an excess capacity in DuPont's waste-handling facilities. Now, customers' waste can be accepted. Another is DuCare®, a developer solvent that does not include TRI chemicals. So, photo developers may be able to reduce emissions below reporting requirement levels and avoid associated expenses.

[14]Exclusive investments include control and remediation technologies; design, project engineering, and installation for abatement equipment; development of Freon® alternatives, and product and process changes for the Montreal Protocol. Largely related investments are those that improve quality or resource efficiency. For example, development and engineering for process improvements that improve first-pass yield. Modestly related are those activities that might improve product quality with some minor environmental improvement.

Leadership and Awareness, Integration, and Vision Contribute to Solving Customer's Environmental Problems

As DuPont changed its perspective on environmental investments from cost-containment to market opportunity, it had to link environmental issues to its business strategy. Integrating these issues with business strategy involved leadership from senior management, greater integration and awareness in business planning, and environmental priorities or vision.

First, senior executive support and involvement was an important contributor to the strategic treatment of environmental R&D. The former chairman of the company, Ed Woolard, chairman from 1989 to 1995, is credited with integrating environmental issues into standard business practices (Schmidheiny, 1992, pp. 193–197; Smart, 1992, p. 187). In the late 1980s and early 1990s the thrust of environmental management at DuPont changed. This change is attributed to Woolard. When he took over the company in 1989, one of his first actions was to identify himself as both the chief executive officer and the chief environmental officer. In his words, "the basic problem was that management values were becoming out of phase with public expectations. Although there were many examples of environmental excellence, they did not reflect a deeply held value of the company." Through a series of organizational changes and his own personal attention, Woolard transformed the company's approach to environmental management from one driven by compliance, to one driven by proactive treatment and business strategy to lead to competitive advantage.[15] His efforts were continued by his successor, Jack Krol, chairman of the board and former president and CEO. An Environmental Business Council, consisting of company business leaders, is a policymaking council that ensures that safety, health, and environmental issues are inserted into the business planning process. This council is supported at the top by the Environmental Policy Committee, a subset of the board of directors. The Director of Safety, Health, and Environment sits on this committee.

Second, in order to influence R&D investments, environmental issues must be integrated with all business decisionmaking. DuPont does this by discussing customers' environmental issues during routine business planning meetings. For example, every 6 to 12 months the vice president of each business area meets with the chairman of DuPont and in attendance is the vice president of Safety, Health, and the Environment. Several environmental topics are addressed during the course of the discussion. The topics include the traditional concerns regarding plant emissions, but they also include

---

[15]For more on Woolard's rationale and actions see Schmidheiny (1992, pp. 193–197), and Smart (1992, pp. 185–202).

discussions about product stewardship and awareness of customer environmental wants and needs. Since evaluating the customer's wants and needs is the primary objective of this process, the vice president of Safety, Health, and Environment might ask: What are your customer's environmental problems, and how are you going to take advantage of this?

This process is similarly performed within the R&D organization as well. The vice president of Environmental Technology will meet with each of the business area's technical directors and laboratory personnel to discuss its program using a template for sustainable business developed by the WBCSD. The WBCSD template categorizes broad sustainability issues as follows:

- minimize toxics dispersion
- maximize recyclability
- minimize energy intensity
- maximize use of renewable resources
- minimize material intensity
- extend product durability
- increase service intensity.

This template has been adapted to conform better to the chemical industry. Applied to DuPont's needs, as driven by its environmental vision statement, the template increases the R&D directors' and managers' awareness of environmental practices and opportunities. This may include anticipating market demands for postconsumer recycling so that the DuPont business area has sufficient time to develop a recycling process; or it may include highlighting the preference for renewable over nonrenewable resources. The primary objectives in the R&D planning process are the same as in the other business planning processes—anticipate customers' environmental wants and needs and anticipate events that may alter the general business position of the company. The business position of DuPont may be altered if new policies that reduce the acceptability of certain chemicals for either DuPont, or its customers, are established (in the past, ODS fit this category), or if input shortages occur that could significantly change prices or availability (such as water or oil). These practices are especially important to environmental R&D in light of the fact that the business areas make the lion's share of R&D investment.

The third element necessary to change its perspective on environmental investments away from cost-containment is priorities or vision. The DuPont Safety, Health, and Environment Vision statement was developed by a group of business, marketing, operating, and technology leaders in a process led by the vice president for Safety, Health, and the Environment. It was sanctioned by the Environmental Business Council. This vision helps communicate environmental priorities to all the business areas, and it is a blueprint for R&D investments. Statements can be literally translated into an R&D agenda. Some of the highlights of this vision are as follows:

- All injuries and occupational illnesses, as well as safety and environmental incidents, are preventable, and our goal for all of them is zero.

- We will assess the environmental impact of each facility we propose to construct and will design, build, operate, and maintain all our facilities and transportation equipment so they are safe and acceptable to local communities and protect the environment.

- We will drive toward zero waste generation at the source. Materials will be reused and recycled to minimize the need for treatment or disposal and to conserve resources.

- We will drive toward zero emissions, giving priority to those that may present the greatest potential risk to health or the environment. Where past practices have created conditions that require correction, we will responsibly correct them.

- We will excel in the efficient use of coal, oil, natural gas, water, minerals, and other natural resources.

- We will extract, make use, handle, package, transport and dispose of our materials safely and in an environmentally responsible manner.

- We will continuously analyze and improve our practices, processes, and products to reduce their risk and impact throughout the product life cycle.

- We will work with our suppliers, carriers, distributors and customers to achieve similar product stewardship, and we will provide information and assistance to support their efforts. (DuPont, 1996a, p. 5.)[16]

Market-driven priorities, or anticipating and meeting the customers' environmental needs are getting serious emphasis from DuPont. However, the vision above includes additional priorities for environmental investments. Investments in R&D to reduce manufacturing emissions can clearly affect DuPont's expenses—in 1994 DuPont spent $765 million on capital equipment and operations for compliance purposes alone (DuPont, 1996a, p. 6).[17] New avoidance or waste-reduction technologies could reduce, or eliminate, the need

---

[16]It took 12 to 18 months to develop this vision.

[17]More recent information was not readily available.

for an estimated $300 million to $500 million worth of additional capital equipment investments as well as operations expenses to comply with environmental regulations. Another estimate, in the 1996 time frame, of potential savings from waste reduction ranged between three billion and five billion dollars for the next 50 percent reduction in waste generated per pound of product.[18]

The Four Categories of Environmental Technologies

Environmental technologies that will be important to DuPont are classified into the following: reuse and recycle, yield improvement (including zero waste and co-product development), reduced emissions technology and abatement, and sustainable products and services.

Depending on the leverage offered by new technologies and the amount of time spent identifying these opportunities, specific priorities from business to business will vary. However, on average these might be, in order: sustainable or environmentally improved products and services; yield improvements including zero waste processes and then zero waste technology; reuse and recycle opportunities; and reduced emissions, treatment, and abatement technologies.

DuPont has historically focused a lot of attention on yield improvements. So, *generally*—recall that DuPont has 17–28 diverse product lines—this area has achieved major advances. There is still room for improvement and even at yields of 95 percent, a one percentage point improvement would be a 20 percent improvement in waste reduction. Within the yield improvement category, process technology changes or material substitutions to reduce or eliminate toxic chemical use and waste are seeing a lot of investment. The second most active area is the development of new input materials that are either renewable or benign. An example is the development of aqueous solutions to substitute for hazardous solvents, such as methylene chloride, or common organic solvents, such as acetone or methylethylketone (Pelley, 1997, pp. 138A–141A). Co-product development can be very high priority or not an issue depending on the specific business and site.

Previous DuPont successes by category are shown in Table C.2.

---

[18]Personal communication with John Carberry (1997) indicated that this number includes material cost savings, disposal cost savings, and lost revenue opportunities. The loss in potential revenues for input materials is the largest portion of this estimate by far. Waste per pound of product will be reduced with the range of DuPont's environmental technology investments—yield improvements, sustainable products, reuse/recycle, zero emissions, and co-product development technologies (DuPont, 1996b).

Table C.2

DuPont's Activities by Environmental Objective

Environmental Objective (percentage)[a]	Purpose/Action	Environmental Outcome	Savings
Co-product development/yield improvement/zero waste technology (40)[b]			
Environmentally superior solvent from nylon intermediate by-products	Generate by-products that can be feedstocks or sold as final products		
Water treatment chemicals from titanium dioxide byproducts	Generate by-products that can be feedstocks or sold as final products		
Beaumont acrylonitrile manufacturing[c]	Process modification to reduce ammonium sulfate waste	Improved yields, reduced ammonium sulfate waste production	Net savings of $1M/year due to resource efficiency and reduced hazardous waste disposal taxes
New Jersey fluorelastomers plant [c]	Process modification to reduce air emissions ($3,000 engineering expense for phase 1, $250,000 for phase 2)	Air emissions cut 50% in phase 1, additional 30% in phase 2	Savings of $400,000/year (alternative to invest in incinerator at $2M capital and $1M/year operations)
Niachlor chlorine plant [d]	Modification to process chemistry, substitute calcium chloride for sodium salt	Reduce solid-waste treatment for sulfates by 95%	$1M/year savings in waste treatment expenses
Lycra manufacturing process[e]	New process that eliminates waste		Nonquantified capital and cost advantages
Reuse and recycle (20)			
Chattanooga nylon fiber plant[d]	Use postconsumer nylon from carpet	Create nylon resins for use with virgin nylon Developing ammonolysis, a process to break down nylon polymer to original monomers	
Tyvek® manufacturing[d]	Meet customer standards for recycled content	Recycle postconsumer waste polyethylene from milk jugs	

Table C.2—Continued

Environmental Objec-tive (percentage)[a]	Purpose/Action	Environmental Outcome	Savings
Reduced emissions and abatement technology (10)			
Conoco Corpus Christi Division[d]	Engineering changes to reduce air emissions	Reduce air emissions by 35%	$4M in permit-related expenses saved
Environmentally preferable products and services (30)			
CFC substitutes[d]	Suva®	Mitigate destruction of ozone	
Agricultural products	Accent®, Londax®, and Classic®	Reduce chemical loading to arable soils	
Printing and Publishing[d]	Waterproof®	Allows use of tap water for color-offset proofing	
Printing and Publishing[d]	DuCare® film processing	Replaces hazardous chemical use with nonhazardous one, allowing for recycling of processing chemicals and eliminating film process effluent from customer's facilities	

SOURCES: [c]Smart, 1992, pp. 185–202. [d]DuPont, 1996a. [e]DuPont, 1996b.
[a]The figure in parentheses is a subjective estimate of environmental R&D distribution
[b]Yield improvement technologies will reduce the opportunities for co-product development and vice versa.

As mentioned earlier, DuPont is beginning to build sustainability into its products and processes. From a business standpoint, the most pressing science and technology needs for sustainability are listed below.

- Understanding what sustainability means in terms that can be operationalized and incorporated into new products and services.

- Determining what constitutes an environmentally benign activity or product.

- Relating dematerialization to business opportunities.

- Performing economical reverse distribution. That is, developing the ability to collect, sort, clean, and disassemble materials at the *molecular* level to be able to make first-quality recycled material.

- Understanding and developing low-cost ways to reduce the energy intensity required to address other environmental issues. For example, develop cheap, lower energy intensity alternatives to materials recycling or zero emissions manufacturing processes.

In summary, DuPont has identified customer-driven environmental investments as its top priority. Following closely behind are process or product technologies to improve yield and eliminate waste. Reuse/recycle and abatement or treatment are other areas receiving some investment. Most of these reflect market-driven priorities. As a result, DuPont has integrated environmental issues into its business planning processes. DuPont is also using sustainability principles, as outlined by the WBCSD, to guide business planning.

PARTNERSHIPS ARE APPROPRIATE FOR CERTAIN CLASSES OF ENVIRONMENTAL TECHNOLOGY

> The competitive advantage of an enterprise is based on technological innovation. Transferring technology means transferring competitive advantage. It is therefore a sensitive issue. Companies want expansion of business through long-term partnerships, not loss of business by the selling of their technologies.
> —*Darwin Wika, manager of environment, health, and safety for DuPont Asia*
> *(Schmidheiny, 1992, pp. 193–197; Smart, 1992, p. 128.)*

DuPont is pleased with the EPA-industry Remediation Technologies Development Forum (RTDF), a public-private partnership. DuPont technology managers noted that this partnership is particularly effective because all stakeholders—industry, government, academia, nongovernmental organizations (NGOs)—are engaged. Each brings an important perspective to the partnership. Industry's role is to focus the technical program. Government provides technology and test opportunities and shepherds the legal issues through the system. Academia provides creative ideas. NGOs and government ensure that public interests are represented, facilitating acceptance.[19] The Chemical Industry Environmental Technology Partnership (CIETP) seeks to develop environmental technologies for the chemical industry. Founded in 1995, the aim of this partnership is to share expenses and risks for environmental technology

---

[19]Remediation technology and service providers, as opposed to the problem-holders, are not a part of this forum. There are at least a couple of reasons for this. This is in part because the forum seeks to address all potential solutions without bias toward a particular approach. Once potential solutions or approaches have been identified, providers may be brought in to contribute advice and expertise. If the forum develops a new technology, it may then be licensed to these providers. Another reason is that the smaller remediation technology and service providers invest little in R&D.

developments that will help interested parties in the chemical industry meet corporate sustainability goals, address environmental regulations, or reduce potential liabilities more cost-effectively. As opposed to the RTDF, CIETP is a more appropriate forum for developing technologies that are more specific to treatment and abatement or can be used for broader chemical processing, where rigorous control of the technology is desired for liability or competitive reasons. A project initiated by some CIETP members seeks to develop technologies to recover chlorine from waste hydrochloric acid. Another project seeks ways to minimize biosludge generation in wastewater treatment plants and to find acceptable use or disposal methods for what is generated. The professional management structure of CIETP provides focused leadership and discipline, an important aspect of this partnership. Another key element is the limitation on liability. For each of these partnership arrangements, technology diffusion may be achieved through licensing.

It is difficult to estimate the total amount of corporate R&D that could be available for partnerships, particularly research consortiums. DuPont's director of Environmental Technology estimates that on the order of 10 to 20 percent of all R&D could potentially be available for research consortiums. Perhaps more of the environmental R&D investments could be involved in partnership arrangements, on the order of 20 to 30 percent. In terms of research and technology areas, partnerships to develop remediation and control are the most feasible or likely. Resource efficiency technologies, for example those that improve first-pass yield, are less likely to be promoted through partnerships. Avoidance technologies are even more competitive and may have to be promoted with other policy tools.

FEDERAL GOVERNMENT SUPPORT FOR THE R&D INFRASTRUCTURE AND ENVIRONMENTAL PRIORITY-SETTING

The director of Environmental Technology at DuPont suggests that the federal government could help promote environmental investments in the following ways:

- Provide incentives to industry to develop environmental technologies such as eco-labeling (based on science), liability reform, and charging customers the full disposal costs.

- Pass environmental regulations that enable pollution prevention rather than those that prescribe behavior. For example, make recycling of wastewater easier.

- Lead and provide a forum for R&D priority-setting with all stakeholders. This process would build consensus among stakeholders, much like the forums that addressed ODS. The forum would also develop a national environmental technology agenda with priorities that could be implemented in industry, universities, and federal laboratories. The difference between this and previous visioning or strategy exercises, which provided a good conceptual overview, is that the concepts would be developed more completely and operationalized into projects, and priorities would be reevaluated on a continuous basis.

- Maintain the R&D infrastructure. That is, focus on the technical strengths of the national laboratories and their ability to synthesize information from universities. Help the laboratories work with industry to screen this information and bench test the promising ones. Support university research as NIH does in health, and make public facilities (federal, state, and local) available as test and demonstration sites. The work Oak Ridge National Lab did with the city of Chattanooga to develop a wastewater technology is a good example.

- Collect and provide data. Scientific knowledge and data on environmental impacts are still needed. For example, data for the environmental impacts portion of life cycle assessments and information on persistent and bio-accumulating chemicals are needed.

SUMMARY

DuPont's view of environmental technology investments has evolved during the previous 10 years. Now, instead of treating these investments as costs driven by environmental regulations, they are treated as opportunities to save money or address markets and, as such, can contribute to the corporation's competitive advantage. DuPont's strategy is to apply its expertise to develop products and services that address its customers' environmental problems, have a positive environmental image, or meet standards and to anticipate changes in the acceptance of its products or input material shortages. To incorporate these strategies into business planning processes, DuPont executives and managers explicitly address customers' environmental needs, sustainability using a template developed by the WBCSD, and DuPont's own Safety, Health, and the Environment Vision.

A large portion of DuPont's R&D, an estimated 95 percent, has some environmental component—although these investments are predominantly made for other primary purposes. Technology categories of primary interest to DuPont are yield improvement/co-product development and sustainable products, followed by recycle/reuse, and control technologies.

DuPont's director of Environmental Technology believes that the federal government could promote environmental technology investments by establishing such incentives as eco-labels and liability reform; enabling pollution prevention within the current regulations; leading forums to develop priorities, which are regularly evaluated; supporting the R&D infrastructure; providing information; and engaging in public-private partnerships to address precompetitive technologies.

INTEL: SURVIVAL OF THE FASTEST IN SEMICONDUCTOR INNOVATION

INTEL IS THE LARGEST SEMICONDUCTOR MANUFACTURER IN THE WORLD

In 1971, Intel introduced the world's first microprocessor, the 4-bit 4004. Today, Intel is the largest microprocessor manufacturer in the world, so large that 80 percent of computers worldwide are running Intel processors (Kirkpatrick, 1997, pp. 60–72). Valued at $100 billion, Intel is almost as large as the Big Three automakers combined (Port, 1996).

Intel annual revenues have grown phenomenally in recent years. Since 1991 annual revenues have climbed from $4.8 billion, more than five-fold to $25.1 billion in 1997 (shown in Figure D.1) (Intel, 1998). To continue this growth, Intel seeks to continue to be the preeminent building-block supplier to the computer industry worldwide. Intel supplies the computer industry with chips, boards, systems and software, which are used by the computer industry as "building blocks" to create advanced personal computing systems. Specific products categories are: processors (microprocessors (CPUs), chipsets, motherboards); networking and communications equipment; semiconductor products (flash memory, embedded control chips). Major customers are computer and peripheral manufacturers, PC users, and other manufacturers (Intel, 1996b, p. 1).

Intel is an international company with more than a dozen major facilities outside of the United States, located in eight countries. There are six sites in the continental United States—in Arizona, California, New Mexico, Oregon, and Washington—with multiple facilities at each and one site in Puerto Rico. Its oldest and smallest manufacturing facility, wafer fabrication facility (fab 4) on the Aloha campus in Oregon, was built in 1978 and was closed in 1996. In recent years Intel has been building a new fabrication facility, an investment of

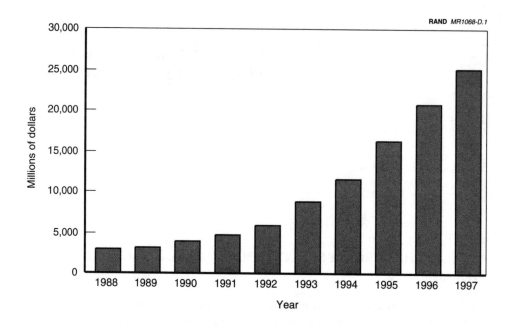

Figure D.1—Intel Revenues

$1 billion to $1.5 billion, every nine to twelve months (Kirkpatrick, 1997, p. 63; McManus, 1996).[1]

The total number of Intel employees was 63,700 at 1997 year-end, up from 48,500 in 1996 (Intel, 1998). Worldwide EH&S staff number around 225 and are organized by site (McManus, 1996; Intel, 1996b). Intel appears to be a relatively flat organization—no product divisions; sales and product groups have a few engineers dedicated to support a specific product. While manufacturing operations are organized under one individual, employees are geographically dispersed. Specific manufacturing project teams are pulled from across the organization as required. Intel has a history of maintaining a fluid and flexible organization where "all such forms are ultimately transitory, and their purpose is to respond to the needs of the time." (Nanda and Bartlett, 1994, p. 4.)

---

[1]It takes between 18 and 22 months to construct a facility.

THE TECHNOLOGY TREADMILL AND SURVIVAL OF THE FASTEST

> We will guard our intellectual property like a hawk, . . . but ultimately, speed is the only weapon we have.
>
> —*Andy Grove, former CEO of Intel (Nanda and Bartlett, 1994, p. 14.)*

> Intel is on a treadmill of new-product introductions fed by increasing demand for microprocessors.
>
> —*Scott Randall, SoundView financial security analyst (Kirkpatrick, 1997, p. 62.)*

A founder and chairman emeritus of Intel, Dr. Gordon Moore is famous for his 1965 prediction that the number of transistors on a chip will double every 18 months (Brandenburger and Nalebuff, 1996, pp. 168–175). This prediction has held true, Intel's primary product, the microprocessor, or chip, is seeing explosive technological change—product complexity has been increasing at an exponential rate. Improvements in two measures of semiconductor complexity, the number of transistors placed on a chip and the number of instructions processed per second, are shown in Figure D.2 below. As shown, the rates of increase in both of these measures have been exponential since introduction of the first semiconductor, the 4004, in 1971. The 4004 chip had 2,300 transistors on it that processed 0.06 million instructions per second. For comparison, the 1286 chip, projected to arrive in 2011, 40 years after the 4004, will have one billion transistors that processes 100,000 million instructions per second.[2]

This rapid series of incremental product improvements is referred to as the Intel "technology treadmill." In the past their business strategy was to innovate faster than competitors, exploit leader advantages and then move on to the next generation of technology. The large profit margins from new semiconductors, usually around 60 percent and as high as 90 percent, fuel further innovation as well as capacity investments in new plant and equipment (Reinhardt et al., 1997, p. 70).[3] According to a recent article, Intel is revising its overall strategy in response to market changes. For the first time it is creating semiconductors for

---

[2]Note: the personal computer was introduced in 1981.

[3]Intel's previous strategy was one of technology push and planned obsolescence. It relied on its ability to bring new technology quickly to the market to drive demand for better and faster computers. Once the technology became mature and competitors moved in, it was ready to introduce the next-generation technology and licensed production of the old to others. For example, in January 1996, Intel signed a technology licensing agreement with Rochester Electronics Inc. (REI), a company that specializes in manufacturing semiconductor products discontinued by the original manufacturers. Under this agreement, REI receives the rights, tooling, and technical data to manufacture 29 mature Intel products to include: 8088 microprocessors, the 82XXX series logic chips, and communication devices, most of which are based on the NMOS process. Intel plans to discontinue in-house manufacturing of these products (Intel, 1996c).

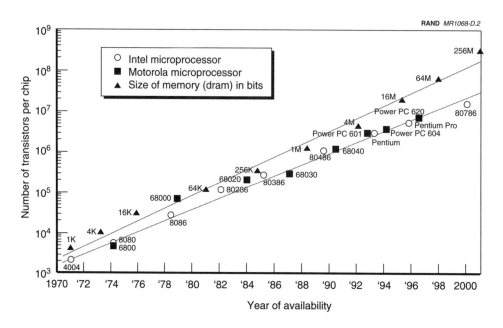

SOURCES: VLSI Research, Inc.; Integrated Circuit Engineering Corp.; Hutcheson and Hutcheson (1996, pp. 54–62).

Figure D.2—Semiconductor Complexity Measure and Performance Improvements

the low-cost computer market, which is expected to grow dramatically in the next several years. In these markets, profit margins are typically lower. This may move Intel to diversify its strategy to address different markets—those that demand rapid product innovation and those that demand low-cost innovations (Reinhardt et al., 1997, pp. 70–77).

On average, new semiconductors take 31 months between generations to develop (Port, 1996, p. 150). In 1970, the R&D investments for a new family of memory chips were $100,000; in 1980, these investments rose to $2 million and on to $100 million in 1985 (Nanda and Bartlett, 1994, p. 12).

Intel seeks these rapid, incremental technological advancements in a very structured development process. They accomplish this in part through concurrent development of product generations and their associated manufacturing processes. That is, new products (including semiconductor chips and software) and their new manufacturing processes are developed simultaneously. However, after a new chip is introduced, the design team does not stop there. Once the microprocessor is developed, the team works on a series of design modifications to miniaturize the feature size and chip size, increasing its processing speed and complexity (see Figure D.3). For example, after the Intel 486 proces-

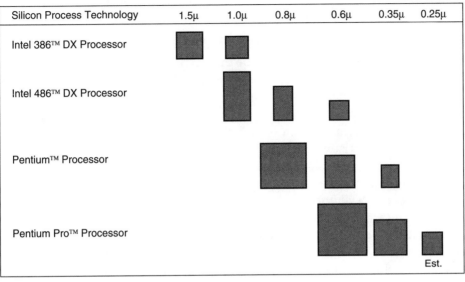

RAND *MR1068-D.3*

Silicon Process Technology	1.5μ	1.0μ	0.8μ	0.6μ	0.35μ	0.25μ
Intel 386™ DX Processor						
Intel 486™ DX Processor						
Pentium™ Processor						
Pentium Pro™ Processor						

SOURCE: Intel Corporation.

Figure D.3—Microprocessor Generations and Feature Size

sor was introduced at the one micron (1.0μ) feature size level, the development team continued working to miniaturize features to the 0.7μ level. These closer features increase the speed of the processor and also increase the number of functions that can be placed on a single chip. These feature sizes are very small. For comparison, bacteria are 8μ in diameter and a human hair is 80μ in diameter.

To accomplish rapid product innovation, Intel also employs two product development teams that "leapfrog" one another to speed the development process. In other words, the two separate product developments are continuously developing new products and when one design team finishes they pick up the follow-on to the other team's chip design. Each chip design team performs a lot of product simulation and verification—including chip layout and functional analyses (co-processor, input/output, etc.).

Another practice employed to reduce the time-to-market is to include manufacturability and customers early in the design process. At the same time product design is underway, manufacturing processes are developed. Approximately one-third of a production line's tool-set is turned over with each transition to a chip with feature sizes at a new micron level. Every two years at Intel, production capability is completely retooled to produce a new generation of semiconductor. Completely new fabrication facilities cost on the order of $1 billion to $1.5 billion dollars and could rise to $10 billion for future facilities (Nanda and Bartlett, 1994, p. 150).

Manufacturing improvements are developed concurrently by a fluid or a virtual organization (Intel employees describe this as "amoeba-like") that supports the entire corporation. These personnel are geographically dispersed and different groups are pulled together on a project basis. After project objectives are met, the organization will restructure to meet the next challenge. Personnel from Hillsboro, Oregon, and Santa Clara, California, generally handle R&D for fabrication, while in Arizona two different groups handle capital equipment planning and chip assembly R&D and fabrication issues. These locations have low-rate, developmental fabrication facilities. At these prove-out facilities, new techniques are developed and tested. Once new processes are sufficiently tested, they are copied exactly (in terms of equipment, procedures, and methodologies) and transferred to the full-rate fabrication facilities. This allows Intel to expand capacity to get out new products quickly, which is especially difficult for chip manufacturing because of the capital intensity of fabrication and the tacit knowledge required to get production yield up. Intel's goal is to make its fabrication facilities identical around the world. In this way, manufacturing operations can be managed more easily and consistently, yields can be kept high, and most important, changes and new techniques can be diffused rapidly. This also gives Intel more flexibility to meet customer demand because identical manufacturing operations will eliminate the customer's need to requalify products from different factories. Finally, the continuous change in products and processes affords Intel many opportunities to at least consider pollution avoidance and other environmentally conscious manufacturing improvements on an incremental basis.

R&D EMPHASIZES DEVELOPMENT

Investments in R&D have grown from $618 million in 1991 to nearly $2.4 billion in 1997 (Intel, 1998). Figure D.4 shows the growth in Intel's R&D investments since 1986. Because much of Intel's competitive advantage derives from its ability to bring new products to market rapidly, most of what Intel does in-house is development work. And for the most part, its development work is specific product development.[4] Intel relies on universities for most of its

---

[4]Intel is not noted for performing original research on semiconductor design. Reduced Instruction Set Computing technology, the basis for the Pentium Pro chip, was originally developed at IBM and Motorola in the late 1980s. Only recently has Intel begun to invest a nominal amount (a few million dollars) into original, long-term computer chip design. This will be performed at its microcomputer lab. Its stated mission is to "keep the technology treadmill going." (*Wall Street Journal*, 1996.)

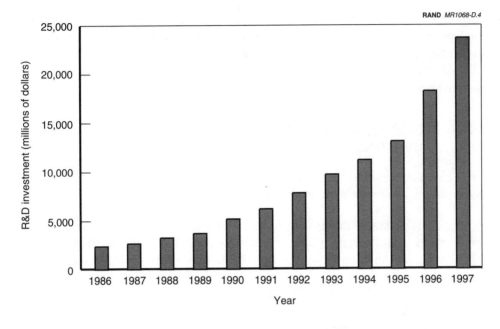

Figure D.4—R&D Investment Growth at Intel

"research," or for projects with a longer-term focus than internal R&D. For example, the EH&S group is sponsoring university research on converting the gas form of xylene to liquid form so that emissions can be managed better.

Intel has a Research Council, consisting of committees organized by their various research labs, to promote and support higher education and research in areas that complement and augment their own programs. Examples of Intel research labs are the components research lab, microcomputer research lab, and applications research lab.

Each university-based project has an Intel sponsor, who is responsible for promoting it within Intel and for ensuring that results are transferred from the university to Intel. In addition to funding university research through grants and gifts, the Research Council can develop mutually supportive relationships in a variety of other ways. These range from participation in Intel research forums and seminars to sponsoring visiting faculty collaboration to one-on-one partnering on projects to hiring graduates.[5]

In order to ensure that environmental issues are integrated with general business-management activities, Intel does not identify environmentally oriented

---

[5]This information is taken from http://www.intel.com/intel/other/rsearch/indesc.htm.

R&D separately. R&D investments are made to meet internal objectives, and environmentally related objectives are a part of the overall set. They are not identified as a separate cost center. One environmental, health, and safety (EH&S) staff member is devoted full-time to the development and execution of the environmentally related objectives.

INTEL IS EXPERIENCED WITH FEDERAL GOVERNMENT COLLABORATIONS

Intel has been, and is, involved in numerous collaborations with the federal government. A sample of their activities follows.

- Intel has several CRADAs with national laboratories. They have several with the Sandia National Laboratory to develop better analytic methods and instruments to sample for one specific solvent when multiple solvents are present in the same tank (McManus, 1997). Another CRADA aims to reduce ultrapure water use during chip rinse (also with a university), and a third aims to develop chemical ionization mass spectroscopy to monitor air emissions with portable equipment. In the past, a Sandia rapid-response team worked with Intel engineers to solve a clogging problem in newly installed thermal oxidizers as well (Intel, 1996b, p. 7 and p. 17).

- Intel has identified a project during the Common Sense Initiative (CSI) implementation to develop better analytic methods to measure removal efficiencies for HAPs when the concentration is low, less than 1 ppm, but flow-through volume is high (such as in a wet scrubber).

- The Semiconductor Industries Association, the EPA, and Intel have signed a memorandum of understanding to identify the best methods to reduce perfluorocarbon emissions (Intel 1996b, p. 9).

- There is an EH&S staff member involved with SEMATECH (full time in 1995). They are developing ultrapure water conservation models and air emissions management systems with SEMATECH (Intel, 1996b, p. 12). However, SEMATECH-funded research is generally secondary to Intel's own research because its development times are just too fast to coordinate research with outside parties. In addition, there are some differences in priorities between Intel, a semiconductor manufacturer, and other industrial members of SEMATECH, most of whom are memory chip producers. Results from this collaborative effort are useful to Intel as alternatives to their own or as a way to refine a basic approach that they've initiated (McManus, 1997).

INTEL SEEKS TO ENGAGE ENVIRONMENTAL POLICYMAKERS

Intel is involved in several voluntary programs. For example, its Aloha, Oregon, facility participated in an experimental air permit program that included pollution-prevention activities. Its Chandler, Arizona, facility is involved in the EPA Project XL Program. As part of Project XL, Intel gains some regulatory flexibility in exchange for agreeing to superior environmental performance and extensive stakeholder involvement in priority-setting. Superior environmental performance is defined by the Project XL program as environmental performance that exceeds what would be achieved through routine compliance to current and reasonably anticipated future regulatory requirements. The specifics of the agreement are documented in an environmental master plan that includes both voluntary components for environmental improvements and two operating permits. This plan was signed November 19, 1996.[6] Extensive stakeholder involvement is a part of priority-setting within this experiment (Intel, EPA, state and local authorities, tribes, and the community).

Intel's primary motivation for these activities is to improve the regulatory process as it relates to its business. Historically, environmental air permits have been issued for five-year increments. However, this is more than two Intel fabrication process "lifetimes." Because it has such a rapid product and process turnover, environmental regulations have become constraining. As a result, permitting times under the Clean Air Act Amendments of 1990 have created a huge opportunity cost for Intel.

At the international level, Intel has worked with the German government to develop the Closed Cycle Substance and Waste Act and the associated Electronic Waste Ordinance. In Israel, Intel worked with the Israeli Ministry of the Environment through the High Technology Industrial Zone to receive and properly dispose of chemical waste generated by the small companies near their facility (Intel, 1996b, p. 7).

Intel managers have suggested that the company pursue these opportunities for a multitude of reasons that range from a corporate emphasis on quality and continuous improvement to good corporate citizenship. According to these

---

[6]The environmental master plan is a comprehensive plan for the facility that includes voluntary actions, such as monitoring; water recycle goals; a single, integrated environmental report; and commitments on setbacks beyond legal requirements. The Intel Project XL final project agreement was signed after nine months of stakeholder discussions. Public comments on the Intel final project agreement were focused on two general areas: the tensions between provisions that provide the flexibility that Intel desires and assurances that superior environmental performance is achieved, maintained, and enforceable; and ensuring that the stakeholder decisionmaking process is as inclusive and equitable as possible. It appears that the lack of toxicological data and risk assessment methods and the inability to closely monitor actual environmental performance, particularly HAP emissions, have contributed to differing interpretations of Intel's plan to achieve superior environmental performance (EPA, 1997; NRDC, 1997).

managers, it is generally taken on faith as opposed to a rigorous calculation of return-on-investment that these activities will have some benefit to the corporate bottom line or to its employees (McManus, 1997; Mohin, 1997).

Intel's Environmental Goals

A large part of environmental management at Intel is employing design-for-environment practices to achieve its own environmental performance goals. These analyses will also help to determine the required investments necessary for achieving desired environmental performance.

Intel is focusing on four areas: air emissions, chemicals use, natural resource use, and solid-waste generation.

Air emissions. Intel's goal is to reduce air emissions to become a minor source at all locations (McManus, 1997; EPA, 1997). As a result, it is working to reduce volatile organic compounds (VOCs) and HAPs emissions through process redesign, chemical reuse and management, natural gas and boiler emissions reductions, and control technology development (Intel, 1996b, p. 6). Eliminating the need for an air permit under Title V of the Clean Air Act Amendments would save the company compliance expenses.[7] Perhaps more valuable to Intel is the time spent negotiating permit modifications. We suspect that given such rapid product and process cycle times any delay can have huge opportunity costs.

Chemical use. Reductions in chemical use are being sought through careful management and Intel's "rent-a-chem" approach (McManus, 1997). They have a Strategic Chemical Council, established in 1993, that reviews chemical use at all sites. This executive-level council reviews products and process designs for improvements in chemical, water, and energy use (Intel, 1996b, p. 8). Ultimately, Intel seeks to close the material loop for chemicals either in an open or closed loop fashion—chemicals can either be completely incorporated into the product, with zero waste, or chemicals can be used, waste can be captured, and returned to the supplier for reuse or recycle. For example, Intel is working with its suppliers to capture PFCs for reuse/recycle and is also working with EPA and the Semiconductors Manufacturing Association to reduce PFC use overall.

Natural resource use. Water is an important issue in semiconductor manufacturing. The millions of gallons of water per day required by semiconductor

---

[7]Potential savings estimates were not provided by Intel. However, information for another semiconductor manufacturer, Lucent Technologies, suggests that the amounts can be large. It makes as many as 50 manufacturing process changes per year at its Allentown, Pennsylvania, semiconductor plant. Almost every one of these changes requires a permit modification, which can take hundreds of thousands of dollars to negotiate (Sanders, 1996, pp. 75–76).

manufacturing generated community concern over local water supplies at its plant in Rio Rancho, New Mexico. An Intel spokesperson was quoted as saying that controversy over water usage at the plant in New Mexico, was "the most intense the industry has faced." (Labate, 1995, p. 32.) At Intel's Chandler, Arizona, facility, which is the one participating in the EPA's Project XL Program, stakeholder discussions uncovered community concerns about water use at this plant as well. While water is not reflected in environmental regulations, its use was of ultimate concern to the community surrounding the plant.

Semiconductor manufacturing requires a lot of water in the various chip cleaning and cooling processes. According to Intel, its Chandler, Arizona, fabrication facility (fabrication facility 12) was estimated to require three to six million gallons of water per day before several water use modifications were introduced. After these modifications are made, water use is expected to drop to one million gallons per day (Intel, 1996b, p. 13).

Two areas have been targeted to reduce water use—chip cleaning and wastewater operations. Chip cleaning operations require ultrapure water. Ultrapure water is required since even with a trace of organics in the water—less than 1 ppm—bacteria could grow and cause plant shutdown for weeks (recall feature sizes are currently in the .7μ to .35μ range while bacteria are 8μ in diameter). Efficient conversion processes require roughly 1.5 to two gallons of potable water to produce one gallon of ultrapure water. Current technological options to the use of ultrapure water for wafer cleaning, such as "no-clean," have been reported as too costly (Labate, 1995, p. 32). Intel's manager for Corporate Environmental Affairs suggests that there are unresolved technical issues as well. Intel is working with its equipment suppliers to develop new wet benches that clean silicon wafers more efficiently and increase ultrapure water recovery. It anticipates that the new benches will reduce ultrapure water use by 40 percent, reducing total water demand by 300,000 gallons per day at each facility (Intel, 1996b, p. 12).

Wastewater is the second area that has been targeted for technological improvement. Intel has developed two models for dealing with wastewater issues, an internal recycle approach (employed at fabrication facility 11 at Rio Rancho, New Mexico) and an external reuse approach (employed at fabrication facility 12 at Chandler, Arizona). Intel will transfer these approaches from its prove-out facilities to its other production facilities, just as it does with general fabrication techniques. The internal water-recycling approach uses process effluent for other purposes, primarily the cooling towers. Since water is internally recycled, management tends to focus on total dissolved solids content, which can accumulate with reuse. In contrast, the external approach involves buying municipal wastewater for use in cooling towers and Xeriscaping. The effluent from chip rinsing and ultrapure water production is treated at a city-facility built specifically for Intel wastewater and funded by Intel. Reverse osmosis is used to treat the wastewater, and then it is injected back into the ground water. Separate lines must be run by both the city and Intel to take in both fresh water and wastewater as well as to discharge treated and untreated water. The external approach is only economically feasible because Chandler, Arizona, is a zero discharge city. That is, Chandler has invested in the infrastructure required to reuse wastewater citywide. Intel personnel could not identify another city that has invested in this infrastructure. Even with the city's preexisting infrastructure investments, Intel had to invest an additional 2 percent, or more than $20 million, more than no-recycling approaches. If it used the no-recycle approach, Intel could have designed its water system to discharge all wastewater to the city for treatment at a cost to Intel of approximately $5 million. Instead, Intel designed its system for external recycling using the city's existing infrastructure. However, Intel spent approximately $28 million for a dedicated water treatment plant that will be owned and operated by the city to process its wastewater by reverse osmosis. Once processed, the water either will be used to replenish the aquifer or will be reused by the water district (McManus, 1997).

According to Intel, the stakeholder process is a useful way to incorporate community priorities, which vary across the nation, into environmental decisionmaking. In the Project XL process, Intel and the community were able to agree upon a mutually beneficial plan—Intel gained the flexibility to make process changes more quickly and the community gained water-use reductions, environmental emissions reductions, one emergency plan, and one consolidated report with all environmentally related information (Mohin, 1996 and 1997).

Solid-waste generation. Intel is addressing solid-waste generation through design-for-environment. These efforts concentrate on reducing and recycling packaging as well as the recycling of scrap silicon wafers generated during fabrication (Intel, 1996b, pp. 10–11).

INTEL'S RECOMMENDATIONS FOR THE FEDERAL ROLE IN ENVIRONMENTAL R&D

Right now regulations send the primary signal to industry about national environmental priorities. From Intel's perspective, regulations have emphasized technology objectives over environmental performance objectives. For example, in the San Francisco Bay area, regulations mandate a 98.5 percent VOC removal efficiency rate. The only way to do this is to use burners, which shift environmental impacts to another form (from VOCs to NO_x and CO_2 emissions because of the increase in natural gas use). In contrast, at another location Intel can use carbon absorbers to mitigate VOC emissions. While these absorbers achieve a 90 percent and not a 98.5 percent removal efficiency rate, they also do not have any associated NO_x and CO_2 emissions. Another example are CFC emissions. These emissions reductions are at the point of diminishing returns—for Intel to eliminate the last 200 pounds of CFCs it may have to invest $5 million to $10 million. At some point cost-effectiveness should be incorporated into regulations.

Intel environmental, health, and safety personnel would also like to see greater allowances for community input into regulatory policy since there are regional variations in priorities and resources. They would also like to ensure that regulations are focused on the appropriate outcomes by improving the way in which rigorous scientific analyses are incorporated into regulatory policy.

Intel environmental personnel would like to see sustainability concepts used for environmental policymaking. In their view, sustainability has a special meaning.[8] It is "when all companies and communities achieve their design-for-environment and these systems are integrated." (McManus, 1997.) They envision a process by which regional or local planning and goal setting will allow companies and communities to establish their own environmental priorities and performance levels. The tricky part of this process then becomes establishing these goals.

Intel personnel's specific recommendations for federal actions to promote environmental R&D are as follows:

- Improve regulatory policy to allow regional approaches and priority-setting (the Dutch model of including communities in priority-setting is a good one) and establish cross-media, performance-based criteria rather than single-medium, technology-based criteria.

- Invest in scientific research and ensure that the new knowledge is incorporated into regulations. Because regulations drive industry's focus it is vital to ensure that they are addressing the right problems. Incorporating new knowledge gains into regulations is just as important as funding the research itself. An area of particular concern to Intel is water quality. In some cases regulations now drive industry management to focus attention on intermediate goals (such as dissolved salt content) rather than the ultimate objective (water use).

---

[8]Note that sustainability is not a well-defined concept. Intel, however, appears to be describing a slightly different process because its description makes no mention of using resources at rates that can be regenerated. However, this could simply be an omission made in the interview process.

- Use federal investments to improve monitoring technologies and measurement standards. Disseminate information and technologies on new analytic techniques and instrumentation. Past experience with EPA suggests that information dissemination and technology diffusion should be given more attention.

- Support the science and technology infrastructure, especially higher education and university-based research.

- Educate the citizenry on environmental issues for two important purposes—one, to generate market demand, and two, so that communities will be better equipped to establish their own environmental goals and targets.

- Promote sustainability and systems thinking to close material loops among all stakeholders.

- Use federal investments to help municipalities provide the necessary infrastructure.

- Intel did not recommend public-private R&D collaborations for its primary technologies, because Intel's technology treadmill is set for a fast pace.

SUMMARY AND OBSERVATIONS

Intel's business strategy is to introduce rapid product innovations more quickly than its competitors. In the past, its business strategy generated large profit margins, which were invested into new product development and manufacturing capability for the next product innovation. This strategy means that every management process, including environmental, must minimize the time-to-market of new products.

Environmental management is geared toward reducing the regulatory burden on time-to-market, primarily by reducing air emissions below permit thresholds. Public comments on the Intel final project agreement for the Project XL experiment suggest areas for environmental R&D. The inability to define superior environmental performance precisely, the lack of toxicological data and risk-assessment methods, and the inability to closely monitor actual environmental performance, particularly HAP emissions, has contributed to differing interpretations regarding Intel's plan to achieve superior environmental performance and its enforceability.

Involvement of stakeholder groups is also a part of Intel's environmental management. The Intel Project XL final project agreement was signed after nine months of stakeholder discussions. Community concerns about water use have led to investments in research and equipment to reduce it.

Regulations are the only existing signal to industry, and, according to Intel, they include single-medium technology objectives not performance objectives. They also do not allow enough flexibility for regional variations in priorities. Intel has its own view of sustainability that it would like to use to plan environmental investments. It is "when all companies and communities achieve their design-for-environment and these systems are integrated." This view includes regional priority-setting to allow for local variability in priorities and stakeholder involvement. Investments in science and technology will ensure that communities have the expertise available to establish scientific environmental priorities. Infrastructure investments will also be required.

R&D collaborations with the federal government have limited application because Intel's technology cycle times are too fast to coordinate with outside entities very well.

MONSANTO: FOOD, HEALTH, HOPE

Headquartered in St. Louis, Missouri, Monsanto is a large, multinational corporation with origins in the chemical business. In 1997, the company split into two parts—Monsanto, a life sciences company, and Solutia, a chemical company. In 1996, sales for the "old" Monsanto exceeded $9.2 billion, more than 40 percent of which were generated in markets outside the United States (Monsanto, 1997a, p. 30). Chemicals (33 percent) and agricultural products (32 percent) generated nearly two-thirds of sales, whereas pharmaceuticals (22 percent) and food ingredients (13 percent) generated the rest (shown in Table E.1).

LONG-TERM RESEARCH COMES HOME TO ROOST: MONSANTO THE STARTUP

Beginning in the late 1970s and continuing largely throughout the 1980s, Monsanto, the chemical company, began long-term research into the biological sciences at its corporate laboratories (Rogers, 1996, pp. 4–9). During the 1980s, a

Table E.1

"Old" Monsanto Sales, R&D Investments, and Employees

Monsanto Prior to Its Split	1994	1995	1996
Total Net Sales (millions)	8,272	8,962	9,262
Agricultural products	2,195	2,441	2,997
Pharmaceuticals	1,520	1,711	1,995
Food ingredients	915	1,117	1,206
Chemicals[a]	3,642	3,693	3,064
R&D investments (millions)[b]	609	658	728
R&D percent of net sales	7.4	7.3	7.9
Employees	29,400	28,500	28,000

SOURCE: Monsanto, 1997a.

[a]1994 and 1995 figures include plastics and rubber chemical businesses.

[b]Computed value based on sales and R&D percentage of sales values.

total of between $800 million to $4 billion was invested in research and acquisitions in biotechnology.[1] Initially, these investments were controversial and
were not supported by all the business sectors. However, as a result of these
investments, a host of new products has begun to emerge, and the company
split for reasons discussed below.

The chemical enterprise, which has taken the name Solutia Inc., includes the
business areas of fibers, chemicals, and polymers and resins. These chemical
products serve a wide range of markets, including apparel, home furnishings,
industrial, automotive, construction, aviation, packaging, and equipment. In
1997, Solutia Inc. had close to $3 billion in sales, 8,800 employees, and 13 man-
ufacturing locations worldwide (see Table E.2) (Solutia, 1998; Monsanto, 1996;
Monsanto 1997d; Monsanto, 1997e). Generally, commodity chemical businesses are capital-intensive and heavily dependent on the price of raw
materials (Schneiderman, 1991, p. 54). Whereas Solutia has retained most of
the chemical products, its business areas cover a spectrum of capital intensity,
in part because some chemicals have higher profit margins.[2]

The life sciences businesses, which maintained the Monsanto name, serve both
commercial and residential markets. Monsanto life sciences includes the business areas of agricultural herbicides; genetically modified cotton, soy, and produce; animal health; food ingredients; and pharmaceuticals. In 1997, it had
$7.5 billion in sales and 21,900 employees, as shown in Table E.2 (Monsanto,
1998). As of this writing, Monsanto life sciences is organizing its business areas
into five sectors: pharmaceuticals, agricultural, health and wellness, nutrition
and consumer products, and sustainable development. The sustainable development sector was established in part to identify market opportunities through
the lens of sustainable development as opposed to that of traditional product
lines. One key element of this sector's strategy is to create market opportunities
from anticipated shortages or constraints in natural resources, such as water
and energy. In general, the life sciences business area seeks to substitute proprietary knowledge for resource content and to market products for their performance, not their composition.[3] Whereas capital investments may be large in
absolute terms, this business area tends to be less capital-intensive (in terms

---

[1]$800 million went to research by one estimate (Schneiderman, 1991, p. 58). $4 billion went to
research and acquisitions by another (Grant, 1997, p. 117).

[2]Capital intensity is defined as the ratio of capital investments required per sales dollar.

[3]For example, chemical products, to a large degree, are a specific combination of other resources—
materials and energy. On the other hand, plants that have been genetically modified to resist pests
are plants with information, not additional material or energy resources.

Table E.2

Monsanto and Solutia Sales, R&D Investments, and Employees

Monsanto Life Sciences	1995	1996	1997
Total net sales (millions)	5,410	6,348	7,514
Agricultural products	2,134	2,555	3,126
Pharmaceuticals	1,711	1,995	2,407
Nutrition and consumer products	1,371	1,581	1,535
Corporate and other	194	217	446
R&D investments (millions)[a]	541	635	902
R&D percent of net sales	10.0	10.0	12.0
Employees			21,900
Solutia			
Total net sales (millions)	2,964	2,977	2,969
Employees			8,800

SOURCES: Monsanto, 1997; Monsanto, 1998; Solutia, 1998.

[a]Computed value based on sales and R&D percentage of sales values.

of investment per dollar of sales) than the chemicals area and is more heavily dependent on R&D investments (NSTC, 1994, p. 54; Brodsky, 1997).

The Monsanto/Solutia split will improve the growth potential of both companies for at least two reasons. First, both companies hope to improve their access to capital, because potential shareholders will be able to more closely match the price-to-earnings ratio with their desired levels of risk. Second, management will be better able to focus its efforts by recognizing and differentiating between the different sets of issues in the two business areas. For example, managers for the more mature product lines in the chemicals company typically are most concerned with cost, quality, and supply issues. As a result, technology management emphasizes cost reduction. In contrast, some of the newer product lines in the Monsanto life sciences company are concerned with creating demand. As a result, the technology issues put product features before cost reduction. And many of Monsanto's investments in the life sciences may take longer to develop into marketable products than will investments in other business areas—on the order of 10 years from an initial idea to product commercialization (largely as a result of the field testing and the regulatory approval processes) (Fritsch, 1996, pp. A3–A4; Monsanto, 1997a, p. 2; Grant, 1997, pp. 116–118). Both Monsanto and Solutia are pursuing diverse product lines—some are established, stable lines, whereas others are new growth areas—so the above characterizations of capital and research intensities may not necessarily apply universally for all products within each of the companies.

The split is the result of a successful, long-term R&D program that led to dramatically different agricultural products, as well as to the expansion into other product areas such as pharmaceuticals. It illustrates how R&D investments can lead to profound change and new markets.

The next section describes how R&D is performed within a large company and provides insights for policymakers who seek to stimulate growth or change. Within this context, the treatment of specialized environmental issues is useful as well.

R&D AT THE COMBINED MONSANTO COMPANY

> R&D strategy addresses predictions about markets and sociopolitical and regulatory changes.
>
> —*Howard Schneiderman (1991, p. 56), former Monsanto senior vice president and chief scientist.*

As noted above, technology-management issues at a large company, such as Monsanto, will vary among the many products offered. R&D for mature products may emphasize cost and reliability of supply issues, whereas new products may emphasize product features, distribution, and marketing.

R&D activities may be characterized in many ways. One descriptor of R&D activity commonly employed is R&D intensity, which is the R&D investment as a percentage of sales. A high intensity suggests that new technologies are required to create future product offerings, which in turn generate sales. Because the "new" Monsanto life sciences company is a technology-based growth company, it is expected that the R&D intensity would be higher than the industrial average.[4]

R&D investments for the combined Monsanto and Solutia companies (the "old" Monsanto), also shown in Table E.1, were between 7 and 8 percent of sales between 1994 and 1996 (According to NSF data the industrial average in 1994 was 3.6 percent.)[5] Investments in R&D for the "old" Monsanto had been rising and were expected to continue to rise as the result of a number of agricultural

---

[4]Monsanto has noted that research intensity is only one descriptor of R&D activity. Since some startup product lines have zero sales, the ratio of investments to sales mathematically approaches infinity. These product lines may not necessarily be research-intensive once sales are established.

[5]Limited data suggest that R&D investments have not always been so significant. In 1979, R&D was less than 3 percent of sales; in 1984, 4.4 percent; 1986, 8 percent; 1989, 7.5 percent. Before Monsanto began investing earnestly in biotechnology, the intensity was in the range of 3 percent of sales and it was focused on process innovation not product innovation. (Schneiderman, 1991; Rogers, 1996, p. 8, and 1997, p. 17.) (For the NSF data, see NSF, 1996, Table A17.)

and pharmaceutical products entering field testing or clinical trials.[6] Therefore, most of this increase was in the life sciences area, which also happens to have higher R&D intensities. Among the more established product lines of the "old" Monsanto, Searle (pharmaceutical products—now part of the "new" Monsanto's pharmaceutical sector) and Ceregen (chemical and biogenetic-based agricultural products—now included within the "new" Monsanto's agricultural sector) had the highest rates of investment in R&D as a percentage of sales (Brodsky, 1997). Historically, pharmaceutical companies typically spent 15 percent of sales on R&D (Schneiderman, 1991, p. 54). In contrast to the life sciences areas, investment in R&D for the chemicals area had remained relatively flat, following the trends for that sector (Brodsky, 1997). These business areas—fibers, plastics, and other performance materials—historically spent 3 percent of sales on R&D whereas commodity chemicals spent less than 3 percent of sales on R&D (Schneiderman, 1991, p. 54).

Detailed information on current R&D investments is not available by business area (although the "new" Monsanto life sciences company reports an R&D intensity of 10 to 12 percent of sales) (Monsanto, 1998, p. 30). However, Monsanto research and technology personnel suggested that these rough percentages still apply (Brodsky, 1997). There is a wide variability of product lines' dependence on advances in technology to maintain or to increase sales even within the same company and within business areas.

R&D Performance

These R&D investments support research at the corporate and business-sector research centers. For example, at the newly formed Monsanto life sciences company, there are one corporate research center and three business-sector research centers (the two business sectors of nutrition and consumer products and sustainable development do not yet have research centers). Each business-sector research center may have specialized skills that relate to its particular set of product lines. The corporate center may have specialized skills geared toward the longer-term research or skills that can be tapped by several business sectors. A management challenge is to exploit the niches of each of the research centers—corporate and business-sector—and to maintain diversity while ensuring that the needs of existing and potential markets are addressed, using all the resources available. The bulk of the R&D investments are made at the newly defined business sectors; less than 5 percent of the total R&D investment is performed at the corporate research center. Where necessary, links are made

---

[6]New drugs may take on the order of 10 years to develop from the laboratory to production (Brodsky, 1997).

to universities, government, and other companies to leverage internal Monsanto research.

Research performed at both the business units and the corporate center relates to future markets but in different ways. The business sectors' centers focus on the more immediate needs of the various product lines that will maintain customer satisfaction and cash flow. The corporate research center performs longer-range, exploratory work that may seek discontinuous change.[7] It may perform research that falls between the dominant markets of the business sectors.[8] Any long-term research must relate in some fashion to a product or market. Some experts label these activities, for lack of a better term, "directed basic research." Note that while the corporate group has more flexibility to pursue particular long-term projects or exploratory research, financial pressures are pushing this work more toward development or shorter-term research.

In general, a technology-oriented or innovative company cannot rely solely on either internal or external research for fundamental scientific advancements. Monsanto has identified numerous acquisitions and alliances to enhance its capability in the life sciences area.[9] When links are made with university researchers to complement the in-house expertise, intellectual property rights are negotiated with the researcher and the university. Specific arrangements may vary. However, the most common approach is for the company to maintain property rights or to have the university maintain the rights with the company having the right of first refusal. Generally speaking, National Science Foundation (NSF) data show that industrial funding of university-based research has been increasing over the past several years. The issue of industrial funding of university research has been of some concern to scientists who feel that negotiating and fulfilling intellectual property agreements can add time and expense to the research process and potentially can limit scientific freedom.[10]

---

[7]Discontinuous or radical change is the opposite of incremental improvements. Typically, a series of small technology improvements are made along a trajectory of performance attributes that have traditionally been used. The core competencies of a company are used to make these improvements. Radical change is a major improvement in a subset, but potentially not all, of these attributes and may draw on new competencies.

[8]Occasionally referred to as the "white spaces," these are potential market opportunities that lie between the dominant markets or the strategies of existing business units. See Byrne (1996), for a discussion of this.

[9]The specific partnerships are briefly described in Monsanto (1997a, pull-out pages) and Monsanto (1998, p. 22), respectively.

[10]For an overview, see W. Wayt Gibbs (1996, pp. 15–16).

Monsanto's Biotechnology Research as an Illustrative Example

The story of Monsanto's transition into biotechnology is a fascinating example of how a company sought discontinuous technological change.[11] In the late 1960s and early 1970s, some scientists at Monsanto's Agricultural Products Company's Research Center recognized that the number of chemical solutions to crop management was limited. They became increasingly interested in biological discoveries at universities for potential applications to both agricultural and pharmaceutical products. However, at the Agricultural Products Company, chemistry, not biology, was the dominant science employed to research and develop future products. Research into the biological sciences would employ an entirely different approach that did not build on internal expertise. Interest in plant growth regulators, chemicals used to modify plant behavior, was high— and many considered alternative approaches to be competitors for funding. Only until the research was brought into the corporate setting in the late 1970s did it truly receive high priority.[12]

Initially, Monsanto tracked biological research advancements (and later cell tissue culture) through a collaboration with Harvard University that began in 1972.[13] This was Monsanto's first major collaboration with a university, and it helped cause "a paradigm shift to biology within Monsanto."[14] This experiment was followed by other collaborations with other universities (e.g., Washington University to study plant transformations) as well as with individual scientists. Later, the company attempted to enter pharmaceutical markets by investing in Genentech, a biotechnology company, and by establishing joint ventures with other biotechnology companies, such as Biogen and Genex. However, eventually, corporate leadership felt that the company had to develop its own expertise and could not rely solely on venture capital.

Several leading scientists were recruited from the universities to establish a program in-house. Experts in insect physiology, plant-cell and tissue culture, microbial genetics, molecular biology, and plant transformation were hired, and a research group was established at the corporate center. The head of this group is quoted as saying "if you can't answer the basic [scientific] questions, you can't have a business." And it was clear that the group's mission was to

---

[11]This is based on Rogers (1996 and 1997).

[12]Schneiderman, former Monsanto senior vice president, Research and Development, recalled in his unpublished autobiography that the corporate research laboratories had been focused for decades on process innovations, as opposed to product innovations, when he arrived at Monsanto in 1979 (Rogers, 1996, p. 8).

[13]The 1972 agreement was Monsanto's first research collaboration with a university (Rogers, 1996, p. 6).

[14]The collaboration began in 1972 and ran through 1984 (Rogers, 1996, p. 6).

advance the science required to "lead to business opportunities in agriculture, animal science, and human health."[15] However, the relevance of this research was not always evident to the business sectors or divisions that were focused on marketable products. For example, when the corporate group sought to develop plant tissue culture techniques that allowed them to grow whole plants from single cells, it used petunias and tobacco because more genetic information was available for these plants at the time. However, the Agricultural Products Company's scientists thought that soybeans, a crop plant, should have been used to develop these techniques. Soybeans, however, had never been grown from a single cell at that time. This would have complicated the research and illustrates the disconnect between those focused on the scientific issues and those more pressured by day-to-day business concerns.

By late 1982, these scientists had made a breakthrough by successfully modifying the genetic makeup of a plant cell. Numerous other fundamental scientific questions were resolved, and, by 1986, work had progressed to the commercial development phase. As a result, the group was transferred to the Agricultural Products Company so that more specific market knowledge could be employed to focus development. By 1987, development had advanced to field testing, which required regulatory approval. Field testing continued and regulatory approvals were sought until the first products were introduced in 1995.[16] These included Roundup Ready® soybeans (soybeans tolerant of Roundup® herbicide), New Leaf® potatoes (potatoes resistant to the Colorado potato beetle), delayed-ripening tomatoes, and Bollgard® cotton (cotton resistant to cotton bollworm, pink bollworm, and tobacco budworm).

Advancements in bioscience and genomics are moving at such a rapid pace that Monsanto hopes to coin a new phrase to draw attention to this change. "Monsanto's Law," fashioned after Moore's Law in the computing industry, suggests that the amount of genetic information used in practical applications will double every year or two (Monsanto, 1998, p. 3). Since the introduction of the initial group of products, Monsanto has sought to rapidly expand its own product realization capability through a variety of means.[17] Through acquisitions, equity investments, and collaborative agreements, it has increased its access to research (e.g., a research agreement with Japan Tobacco Inc. aims to

---

[15]Ernie Jaworski, former manager of the corporate Molecular Biology Group, quoted in Rogers (1997, p. 15).

[16]A new drug or a major crop chemical generally takes about 10 years to get from concept to commercialization, at an investment of $125 million to $200 million. Although many products can take this long as well, attaining regulatory approvals particularly drives this time and expense (Schneiderman, 1991, p. 53).

[17]Product realization refers to the entire set of activities from research to production to product delivery.

develop an improved rice for the Japanese market, and a collaborative agreement with Millennium Pharmaceuticals exists for genomics research), technology (e.g., the acquisition of Biopol provided the genes and related technologies for developing biodegradable plastic), production capability (e.g., the acquisition of Holden's Foundation Seeds Inc. allows Monsanto to introduce genetically modified crops more quickly to seed companies), and distribution channels (e.g., the acquisition of Monsoy provides access to the Brazilian soybean market, which is the second largest in the world).[18]

These strategies illustrate how one corporation sought discontinuous change through R&D investments and collaborations. This change was initiated in part because of poor sales growth in the late 1960s, which led to the realization by some scientists that chemical solutions to crop management had limited growth potential. Biotechnology, an alternative approach, had long-term potential. Nearly 25 years after its initial investment in biology,[19] more than 40 years after J. D. Watson and F. H. C. Crick described deoxyribonucleic acid (DNA), and 139 years after Gregor Mendel began his experiments with peas, Monsanto has introduced several dramatically new biotechnology-based products and opened many other potential markets.

Partnerships

Partnerships in general have been and will continue to be very important to all innovative companies, especially those in scientific fields undergoing rapid change, such as biotechnology. The many dimensions of change in this area—genomic research, plant and animal cloning, and marketing, to name a few—make it unlikely that any single company will gain and apply expertise quickly enough on its own, without partnerships. These unions can take many forms—alliances, collaborations, acquisitions—and may extend beyond research collaborations to production, distribution, and other activities. According to Monsanto Vice President for Research and Environmental Technology Phil Brodsky (1997), the advantage of partnerships is that products and technologies get to market faster; the disadvantage is that part of the market is compromised a priori. The overall proportion of research funding that may be applied to partnerships is specific to the company and the technology involved, and no quantitative estimates of the range of these proportions were provided.

---

[18]A complete listing of acquisitions and alliances by business area is provided in Monsanto (1997, foldout section) and in Monsanto (1998, pp. 22–23).

[19]The nearly 25 years begins with the Harvard University agreement made in 1972 to 1995, when the first products (Roundup Ready® soybeans, NewLeaf® potatoes, delayed ripening tomatoes, and Bollgard® insect-protected cotton) were approved by the U.S. government.

One type of partnership of particular interest to policymakers is the public-private partnership. Generally speaking, research managers at Monsanto see public-private partnerships as useful policy tools for precompetitive or generic research. In addition, companies employ them to seek secondary or alternative technological approaches to existing options. For example, Monsanto is involved in a partnership with the Department of Defense, Teledyne, and Phillips Petroleum to develop a field-deployable mass spectrometer. This technology would offer improvements over current practice, specifically real-time measurement capability, but these improvements are not essential for Monsanto's operations. Another beneficial area for public-private partnerships is in sharing the expense of the research infrastructure. For example, Monsanto is partnered with the Department of Energy and other pharmaceutical companies to invest in a synchrotron for beam-line high-resolution crystallography at Argonne National Laboratory. The investment in the synchrotron, which provides the necessary light source, is then shared, and companies protect the competitive aspects of the research by providing their own crystallized enzymes to develop new drug therapies.

In summary, R&D issues are wide-ranging according to the specific technology and product at issue. Such parameters as the investment intensity, the length of time from the initial idea to market introduction, the focus of research (i.e., on product features as opposed to cost reduction or process improvements), and the types and kinds of partnerships may vary. Although all research is directed toward anticipated markets, research centers at the corporate level generally have more flexibility and more of a longer-range focus than do business sector centers. However, the amount of funding invested at the corporate centers is much lower overall. Finally, companies will employ a variety of legal and organizational mechanisms to leverage internal expertise and capacity with capabilities at universities, governments, and other firms.

The Environmental Challenge Grant Program Brought Additional Creativity to Address Monsanto's Needs

The environmental challenge grant was a novel approach initiated by Earl Beaver, director, Waste Minimization, to find a solution to a vexing problem for Monsanto. The program represents an interesting mechanism for identifying creative technological solutions to a problem.[20]

---

[20]The specific application in this example is no longer of interest to Monsanto the life sciences company, it now belongs to Solutia.

Toxics Release Inventory (TRI) information[21] and emissions analyses indicated that ammonium sulfate made up approximately 76 percent of Monsanto's waste stream as defined by reporting requirements. Ammonia and organics were typically present in these waste streams. If the ammonia and organics could be separated, each could be more easily treated or reused (Beaver, 1996). This in turn would reduce these waste streams as reported in the TRI and therefore could improve the company's environmental reputation. And yet, for nearly 30 years, Monsanto scientists and engineers could not come up with cost-effective ways to reduce or eliminate this waste stream.

In an effort to find a solution to this problem, Monsanto decided to bring additional creativity to bear and to "challenge" others outside the company to propose and develop a viable approach. In return, Monsanto would apply some of the money that was in effect wasted by this discharge and agreed to pay up to $500,000 for additional development work to demonstrate in the laboratory that the approach could be cost-effective and commercially viable. An additional $500,000 would be provided to the developers if the demonstration yielded a successful approach. In return, Monsanto would have a solution to a vexing problem and would hold the property rights to the new technology. In the meantime, Monsanto's own scientists and engineers could focus on reducing toxic chemicals at their source (Beaver, 1996).

The first challenge was issued in August 1994. An independent panel assembled by the Center for Waste Reduction Technologies (CWRT)[22] reviewed the proposals to protect proprietary information. Based on this review, Monsanto selected SRI, International, to develop its technology and to demonstrate commercial feasibility at bench-scale. After further development and subsequent modification of the proposed approach, SRI, International, has been awarded the second $500,000 installment for the development of a commercially promising recovery technology. This waste stream belongs to the chemical company side of Monsanto that now bears the name Solutia, which has filed a patent application now pending. According to the Director for Waste Minimization and initiator of the Challenge Grant Program, plans for ultimate implementation by Solutia are unknown and may have yet to be determined (the project will have to compete with other proposals for capital funds) (Beaver, 1997).

---

[21]The Toxic Release Inventory (TRI) is a requirement under Superfund Amendments and Reauthorization Act (SARA), Title III. Beginning in 1988, it requires companies that meet certain threshold criteria to report emissions of toxic chemicals to air, land, and soil.

[22]The Center for Waste Reduction Technologies is an affiliate of the American Institute of Chemical Engineers. Go to http://www.aiche.org/docs/cwrt/ for more information.

The second challenge issued by Monsanto (and transferred to Solutia as well) sought technologies to remove sodium chloride, a high-purity amino acid, and phosphoric and phosphorous acids from wastewater streams for potential reuse or recycle (Monsanto, 1995). Also won by SRI, International, this project is in development to demonstrate commercial feasibility. Earl Beaver (1995) of Monsanto described the challenges program:

> We'd like to think our Challenges are serving a purpose beyond just Monsanto's needs. . . . We hope they are causing people in many quarters—in other industries, in the research community, and in government—to think more broadly, more creatively, perhaps more cooperatively than they have in the past about how to accelerate environmental solutions.

This challenge approach has been used twice now and is considered a success. In addition to the benefit to Monsanto and now Solutia, some proposals led to collaborations with the CWRT itself.

The challenge concept is a novel approach, and managing the process can be difficult and time consuming. For example, Monsanto had to pursue extensive outreach to advertise the program worldwide—since its goal was to capture a large applicant pool and maximize the creativity that would be applied to its problem. Because Monsanto had a lot of experience with ammonium sulfate waste issues it was well-positioned to specify requirements as well as to evaluate proposals. Yet it had to rely on a third party, the CWRT, to review the proposals to protect their confidentiality and proprietary aspects. As a result, the decision was made with very little direct information. For more exploratory research, this approach may be more difficult to employ.

In summary, the challenge program is one approach Monsanto took to tap outside expertise to solve its problems. In terms of its specific technology applications, these waste streams "belong" to Solutia and as such are not necessarily representative examples of Monsanto's current concerns. Monsanto personnel expressed a willingness to partner, outsource, or look externally for technology solutions to environmental issues that *do not* involve source reduction. In contrast, source reduction research is generally performed in-house.

IDENTIFYING ENVIRONMENTAL R&D IS A DUBIOUS ENDEAVOR

Of particular interest to environmental policymakers is how Monsanto treats R&D to meet environmental objectives. Monsanto does not track environmental R&D investments in the broadest sense. Only a small subset of the research with environmental implications is identified as such for planning purposes. A large amount of research is called environmental only after the fact, when it is discovered that environmental benefits exist.

For planning purposes at Monsanto, environmental R&D has been narrowly defined and is consequently small. It is defined as the R&D performed to meet regulations or to address issues raised by environmental groups. Because these are the primary motivators, such investments generally do not meet corporate financial criteria for R&D investments. Often this work is not performed within Monsanto's research centers and is contracted out. Research into emissions control devices or remediation technologies are obvious examples of research that falls into this category.

However, a lot of R&D performed might not be considered environmental ex ante. These investments are generally driven by the desire to improve production yield, to decrease waste generation, to reduce costs, or to develop new products (with improved environmental attributes). They meet corporate financial criteria. This R&D is often called environmental R&D after the fact because of some positive effect on environmental performance. For example, an analysis of the costs associated with the manufacture of a rubber stabilizer showed that money was being wasted. This particular manufacturing process employed standard industry practices, and the chemistry was considered mature and efficient. However, the cost analysis showed that 1.5 pounds of waste was generated per pound of product—money lost to Monsanto. The corporate research center invested some R&D money to explore alternative manufacturing techniques, and a new chemistry was discovered. The new process reduced waste generated to 0.1 pound per pound of product. While this effort was categorized as a cost-reduction effort, there were clear environmental benefits as well. Another example is Monsanto's Roundup® herbicide. Because this herbicide degrades rapidly in the field, it could be considered environmentally preferable to previous herbicides. Only after Roundup® had been used for some time was it discovered that a real environmental benefit was that farmers could avoid tilling fields. This benefits the environment because no-till reduces soil erosion, energy use, and CO_2 emissions.[23] However, investments to develop Roundup® or Roundup Ready® soybeans, which are soybeans resistant to the herbicide, were not considered environmental R&D. In many of these cases it would be very difficult to assign the portion of the R&D investment driven by environmental implications rather than by product or process improvement, either ex ante or ex post.

While environmental research at Monsanto is a bit of a misnomer, in the past relatively little R&D was *driven* by these concerns. A large portion of what could eventually be called environmental R&D was driven by other objectives. It was performed because environmental objectives were consistent with other corpo-

---

[23]CO_2 emissions are reduced because energy use is reduced and because the carbon in the plants and soil is sequestered.

rate objectives of cost reduction, waste reduction, and customer interests. In the past, environmental implications were not an explicit planning criterion. This may change as the concept of sustainability becomes more central to Monsanto's technology development and business practices. This suggests that such traditional tools as environmental regulations can and do influence corporate investments in new technology. However, policies that influence markets and the price of inputs and wastes will have a more profound effect on these investments because they influence a larger portion of them.

THE TRANSITION FROM COMPLIANCE TO BEYOND COMPLIANCE TO SUSTAINABILITY

Monsanto's treatment of environmental issues, which affects its demand for new technology, has evolved over time. This transformation came gradually in response to changing public interests, regulations, market forces, and business strategy and interests. Beginning in 1988, the TRI reporting made company leaders and managers painfully aware of the company's emissions of the listed pollutants and of the associated economic and social implications of these emissions. At that time, company management decided to announce a new effort to reduce hazardous air emissions 90 percent by 1992 from a 1987 baseline. This marked a turning point in the treatment of environmental issues from meeting regulatory requirements to thinking proactively—or beyond compliance requirements. The change in thinking began to permeate the organization as the company actively sought to reduce emissions, using technology and techniques that it found appropriate (Smart, 1992, pp. 123–124). This increased awareness, and the success of the air program, led to Monsanto's waste-reduction program.

Initiated in 1990, the waste-reduction program sought a 70 percent reduction in the TRI chemicals emitted to all media by 1995 (Beaver, 1997). In the end, the program goal was not met—a 33 percent reduction was achieved. As Monsanto's CEO explained in the *1996 Environmental Annual Review,* the only options to reduce toxic emissions were expensive. And they were not "sustainable" solutions—that is, they reduced toxic emissions but actually increased overall waste emissions (Monsanto, 1997b, p. 1). Today, continuous reduction of waste emissions is sought (Beaver, 1998).

After a 1995 meeting of Monsanto personnel called Global Forum, another milestone in the treatment of environmental issues was passed. Sustainability was elevated to a strategic business issue. Sustainability extends the concept of toxic waste minimization to overall resource efficiency. It focuses the company on markets or products that improve society's quality of life and resource efficiency. Therefore, it is a much broader concept than environmental protection.

It takes a worldwide view of markets and issues and seeks to meet these futures. The life sciences company in particular is more closely linked with major environmental issues, such as global warming, land use, soil erosion, and water quality (Hamilton, 1997, p. H1; Grant, 1997, p. 116).

Monsanto CEO Robert Shapiro has articulated the solution to population growth, the major driver in environmental harm, as raising the standard of living around the world. Businesses that help raise living standards—such as those in health, food, energy, and water—will lead to economic growth and improved quality of life and environment (in part by easing the pressure of population growth). Improvements in living standards will in turn increase the demand for Monsanto's products—which is a priority for stockholders.[24] As a result, environmental goals and strategy are being integrated with other corporate goals and strategy.

Sustainability Leads to Business Opportunities

> Even if we choose the minimum goal of maintaining the status quo, we must get twice the yield from every acre of land just to maintain current levels of poverty and malnutrition. . . . The earth can't withstand a systematic increase of material things. If we grow by using more stuff, I'm afraid we'd better start looking for a new planet. . . . Infotech is going to be our most powerful tool. It will let us miniaturize things, avoid waste, and produce more value without producing more stuff.
>
> —*Robert Shapiro, CEO Monsanto (Grant, 1997, p. 117).*

Since Monsanto anticipates a future where resources become constraints—as both population and world economies grow—it is introducing the concept of sustainability into its plans for future products and services. At its core, sustainability is about finding opportunities by decoupling economic growth from increased use of material and energy resources (Monsanto, 1998, p. 2).

Sustainability is operationalized at Monsanto as the process of doing more with less.[25] It is therefore a process, or a way of viewing market opportunities, as opposed to an endpoint or a goal. While there are strong links to environmental issues, sustainability is not viewed as an environmental strategy so much as a standard business strategy.

---

[24]As a multinational, Monsanto could also benefit from the security side of sustainability—that access to markets and resources will be more stable and reliable.

[25]This is how the WBCSD defines eco-efficiency (Fussler with James, 1996, p. 351).

The central sustainability strategy is to develop and provide products and services that add value to either the company or its customers. Value can come in the form of reduced cost—because of improved resource utilization—or added functionality without added raw materials or energy. Thus, as resources become constrained, new markets for resource efficient products will be created. Moreover, as living standards rise, the demand for such products and services will increase. And finally, the destabilizing effects of unsustainable activities add risk and uncertainty to the international operations of large companies. For these reasons the sustainability strategy is anticipated to be "win-win," and the key question then becomes how to anticipate market discontinuities and time investments accordingly.

Underlying this business strategy is a reliance on new technology, particularly biotechnology, which is seen as a subset of information technology. "Information technology is going to be our most powerful tool. It will let us miniaturize things, avoid waste, and produce more value without producing and processing more stuff." (Shapiro, 1997, p. 83.) For instance, information encoded in seeds can generate plants that produce more product or value per unit input. Genetic material can be added to develop plants that require less water, energy, or pesticides. Another example of producing more value without using more material resources is plants that yield both food products and other products, such as biodegradable plastic from the nonedible portions (leaves). Another example would be using recipes for Roundup® herbicide that take differing weather conditions into consideration could lead to the same level of weed removal with smaller quantities of herbicide. Specific patient information, such as weight, age, and sex, coupled with drug therapies, could potentially result in more effective treatments, with smaller dosages. All of these point out how the addition of information, provided in part by advancements in science and technology, can lead to greater material and energy resource efficiencies.

Organizational Change Is Under Way

> The more any business understands about its resource use and impacts of its products, the more profitable it can become. . . . The more we look at the metric, the more opportunities we discover.
>
> —*Paula Menten, leader of the Sustainability Index Team (Harpole, 1997, p. 121).*

In 1996 several teams were created to help focus Monsanto on sustainability by monitoring issues of strategic importance to the company. These teams began to monitor long-term trends in the global drivers of sustainability—social, economic, and environmental issues—and look strategically for business opportunities. There were teams to track global water issues,[26] world hunger issues,[27] and new business opportunities.[28] A few other teams began to develop the analysis tools required to implement the concept of sustainability into business processes and decisions. These included an eco-efficiency[29] team to track material and energy flows for Monsanto's industrial processes;[30] a sustainability matrix team, which developed a product-oriented life-cycle tool using qualitative and quantitative social, environmental, and economic indicators; and an accounting team to develop full cost accounting capability.[31] Without comprehensive analysis tools and good information, sustainability thinking will not be easily integrated into business practices.

It is worth noting that the analysis tools are still in development. Monsanto has found that available tools were not as complete or practical to apply for its purposes. In fact, continuation of the analytic tool development represents two of Monsanto's five sustainability goals. By the end of 1998 it seeks to (1) develop a plan for creating and integrating sustainable development criteria and goals into all key business decisionmaking processes and (2) define these criteria for all Monsanto products with a plan to apply them.[32] Toward this end, Monsanto is collaborating with several organizations to develop available tools. It is participating in a CWRT project with 12 other companies to develop full cost accounting techniques that can more readily be used by several process industries—pharmaceuticals, pulp and paper, and chemicals.[33] Eco-efficiency indicators and tools are being discussed or developed by several other orga-

---

[26]For example, industrial water treatment, potable water availability, irrigation water use and soil salinity, etc., as well as technologies that could be employed to meet these needs, were tracked.

[27]This team looked into ways that the life sciences company might effectively leverage its resources to alleviate the causes of world hunger.

[28]This team took a shorter-term business view of the strategic issues of water and world hunger and attempted to find technologies or products to meet these needs. Based on work in 1995 and 1996, this unit has identified such projects as the development of bioengineered plants that can be used to produce biodegradable plastics or polymers and home and garden plants that resist disease and insects as projects to pursue.

[29]Eco-efficiency is defined in Fiksel (1996, p. 499) as, "The ability of a managed entity to simultaneously meet cost, quality, and performance goals, reduce environmental impacts, and conserve valuable resources."

[30]This life-cycle inventory tool incorporates flows both "upstream" and "downstream" from the Monsanto facility and incorporates a broader view of wastes. Where toxics were traditionally considered in the past, the eco-efficiency tool includes all resources in the input and waste streams, such as water and CO_2.

[31]Full cost accounting includes the total lifetime environmental costs to the product or process that generates them, including such intangibles as public and customer relations.

[32]The others are (3) encourage the workforce to learn about sustainable development; (4) create a plan for developing human and social capital in emerging markets; and (5) create a plan for incorporating external stakeholder inputs and for internal communications. (Monsanto, 1998b, p. 6.)

[33]For more on the project scope and status, see the CWRT website at http://www.aiche.org/docs/cwrt.

nizations. These include CWRT, WBCSD, and the National Roundtable on the Environment and Economy.[34]

In contrast to tool development and data collection, actually applying each of these tools will require an in-depth knowledge of Monsanto products and operations. These tools have been, and will be, used to evaluate projects up for review and to guide corporate investment decisions (Monsanto, 1997c, pp. 20–23; Brodsky and Beaver, 1997). At this writing, the tools have been used to guide decisions regarding new product development, but it is too early for them to have affected investment decisions just yet (Beaver, 1997). These analytic tools are significant in that few companies explicitly integrate sustainability into their management processes to this degree.[35] Those that have integrated sustainability thinking into their management processes have found that anticipatory and preventive actions save money—and in the end are simply good business practices, as suggested by the quotation at the beginning of this subsection.[36]

In 1997 a new sustainable development business sector was established to identify and address emerging markets. The sustainable development sector contains several product lines, including the Enviro-Chem products for air pollution control, odor control and biofiltration; a portable water treatment system; and a biodegradable plastic material. It has also identified the areas of land use, fresh water availability, and home and garden "smart plants" for exploration in the longer term. The work of the sustainability teams described earlier has largely been incorporated into the sustainable development business sector,[37] although some activities are dispersed throughout the corporation.

In summary, the treatment of environmental issues has evolved over the last 10 years from compliance to the broader view of sustainability. Sustainability at its core is operationalized at Monsanto by incorporating resource efficiency while creating value for customers and shareholders (or eco-efficiency). It is a more integrated view of economic, social and environmental issues—the three bot-

---

[34]For more on these activities see the URL in the previous footnote and http://www.nrtee-trnee.ca and http://www.wbcsd.ch/.

[35]This is not to say that no other companies are incorporating this thinking into their management processes—Volvo, Xerox, Dow-Europe, and Nortel are a few who are. The companies involved in the collaborative analysis tool developments are also incorporating that thinking. Moreover, as experience with the International Standardization Organization Environment Management Standards, ISO 14000, grows, others will as well.

[36]The management practices included in Total Quality Management (TQM) seek to gain greater understanding and control of business practices so that the company can optimize these processes and knows exactly how and when to meet changing markets.

[37]It has a center of excellence, which is where most of the education and metrics development work is continuing.

tom lines—and their effects on quality of life than has been employed in the past. Monsanto is developing and deploying business analysis tools and information to identify market opportunities, to measure progress toward sustainability, and to communicate its actions to the public. If successful, this will mean that environmental issues are fully integrated into corporate strategy and planning. These analysis tools are also very important because they contain the information and the measures that will affect corporate behavior and decisions on capital investments, R&D investments, and new product development. It is crucial that the measures and the indicators accurately portray proper social, environmental, and economic objectives.

Advances in Biotechnology Are Not Without Controversy

> The biggest problems that many biotechnology-based businesses face today are social, not scientific, particularly in the application of biotechnology to agriculture.
>
> —*Howard Schneiderman, former Monsanto senior vice president and chief scientist (Schneiderman, 1991, p. 57).*

> [A]ny powerful new technology is going to create ethical problems—problems of privacy, fairness, ethics, power, or control. With any major change in the technological substrate, society has to solve those inherent issues.
>
> —*Robert Shapiro, CEO Monsanto (Magretta, 1997, p. 83).*

A large part of the sustainability strategy of the life sciences company involves biotechnology. For example, genetically modified agricultural products can potentially offer many benefits to consumers, producers, and the environment. However, as with many new technologies, there are risks and uncertainties. Some consumer and environmental groups are uneasy about genetically modified agricultural products for several reasons. These include the possibilities that:

- genetically modified foods may make people ill because genes may be transferred from a plant that was not meant to be consumed;[38]

- modified plants may become uncontrollable weeds (*Economist*, 1997, pp. 80–82; Kling, 1996, pp. 180–181);

---

[38]A Monsanto spokesperson mentioned that their observation is that illness is most often discussed in the context of an allergic reaction.

- modified plants may encourage resistance in pests through natural selection (*Economist*, 1997, pp. 80–82; Grant, 1997, p. 118; Woodfin, 1997, p. 27);

- the marker genes resistant to antibiotics (used to create genetically modified plants) may mingle with bacteria to create a new form of bacteria resistant to antibiotics (*Economist*, 1997, pp. 80–82); and

- transgenic varieties could introduce a dominate genotype into the wild and ultimately reduce the plant's gene pool (Kling, 1996, pp. 180–181).

Most of these issues may be true for plants and pesticides in general, whether or not they've been genetically modified (*Economist*, 1997, p. 80). The fourth item above, while possible in theory, has not been observed in experiments (*Economist*, 1997, p. 80).

With rigorous testing, such mitigating measures as refuge areas, and strict regulatory oversight, the additional risk of genetically modified plants is small. Ultimately society must decide if the benefits of these new technologies— reduced world hunger and soil erosion, reduced water use, etc.—outweigh the potential risks—biodiversity loss, resistant weeds, etc. This highlights the importance of having scientific knowledge, sound regulatory institutions, and good consumer awareness to make these decisions.

RECOMMENDATIONS FOR FEDERAL POLICIES TO IMPROVE INVESTMENTS IN ENVIRONMENTAL TECHNOLOGIES

Monsanto environmental R&D and waste elimination experts would like to see the federal government pursue stimulation or promotion of R&D investments in environmental technologies in several areas. These are (in no particular order) as follows:

- Improve public education and awareness on scientific and technical issues so that product and process innovations will be more readily understood and accepted. Monsanto is especially concerned with the public's reaction to biotechnology products.

- Ensure that regulatory processes at the EPA, FDA, and USDA are rigorous and trusted by the public. Increase funding to these agencies so that they can hire additional, high-quality scientists to assess new products in a timely manner.

- Work internationally to ensure global enforcement of patent laws to protect intellectual property.

- Develop decision and analysis tools that industry can use to find sustainable practices and products. Examples include sustainable product design guidelines, decision matrices, and full cost accounting techniques.

- Continue public-private partnerships and increase cost-sharing to develop generic, precompetitive technologies. Monsanto's priority areas are resource efficiency, waste elimination, and avoidance technologies. To satisfy the criterion that partnerships develop precompetitive technologies, most candidate technologies are likely to be process-oriented and not product-oriented. While environmental remediation technologies tend to be generic technologies, they are not a priority for Monsanto. Partnerships to share scientific or technical infrastructure are also of interest to Monsanto.

- Increase support for basic chemical and biological research at universities, especially chemical research based on biological analogies. Biological analogies are especially interesting because these processes are benign. Support such technology areas as biotechnology, information technology, and nanotechnology.

- Eliminate barriers to seeking new technology internationally. Limiting federally sponsored research to U.S.-based activities reduces the breadth of innovative approaches examined and slows technology diffusion.

- Develop and establish national verification and certification processes for innovative environmental technologies to reduce the time and expense currently required to satisfy differing state standards and requirements.

- Harmonize and simplify the processes as well as the contractual procedures required for performing interagency-funded work.

- Continue and expand positive-oriented activities, such as the EPA's Green Chemistry Program, that reward success.

LESSONS LEARNED

Large corporations will seek discontinuous technological change driven by expectations of future market growth, even outside of their traditional product lines. Monsanto has done this by leveraging its extensive R&D capabilities. While internal R&D is the preeminent contributor to a company's technological expertise, companies expand their technical expertise through a variety of means—through partnerships with universities, alliances, acquisitions, etc.

The amount of environmental R&D is difficult to quantify because in many cases environmental concerns are only one of the many factors that influence investments. At Monsanto, the traditional view of environmental invest-

ments—whatever is necessary to meet regulatory requirements—is small rel-
ative to the rest of the portfolio. The largest portion of R&D is geared toward
products and processes, which have profound environmental implications.
Therefore, estimates of "environmental R&D" might not provide meaningful
information for policymakers when formulating policies to promote beneficial
environmental behavior.

Drivers of R&D with environmental implications are resource efficiency, cost
reduction, waste reduction, customer or market priorities, regulations, and
public concerns. The largest of these investments at Monsanto fall into the resource efficiency, cost reduction, and market priorities categories. They meet
corporate financial criteria and as such are fiscally sound investments. So,
environmental regulations can be a forcing function for technological
investments and may be appropriate when very specific kinds of investments
are desired. Since resource efficiency receives the lion's share of attention,
fiscal and other policies that increase the cost of inputs and wastes may provide
the most flexible and effective tools in some cases. These in turn will promote
research investments to develop the technology to meet markets.

Monsanto has embraced sustainability, or the journey to resource efficiency, as
its strategic direction. By employing the broad scientific and technological
areas of biology, genetics, and computers, the company hopes to increase value
or functionality at lower resource use. It has forged relationships with several
organizations to advance and operationalize the broad concept of sustainability.

Many of the recommendations for federal action were for policies and investments to promote research and to maintain the scientific infrastructure. Policies to increase consumer awareness, support for universities, and rigorous
regulatory processes are examples. Public-private partnerships in generic
resource efficiency or waste-minimization technologies are of interest to Monsanto as are public-private partnerships for the scientific research infrastructure.

XEROX: THE DOCUMENT COMPANY

CORPORATE OVERVIEW OF XEROX

Broadly speaking, Xerox's primary business area is document management, which means that its products and services are devised to facilitate the creation, reproduction, distribution, and storage of documents. Major product lines include publishing systems, copiers, printers, scanners, fax machines, and document management software, along with related supplies and services. Its logo, a stylized "X," represents the pixels of digital imaging, the foundation of the second office revolution.

There are three geographic customer operations units and nine document-processing business divisions. This organizational structure has existed since 1992, when changes were made to align technology development more closely with customer interests. The resulting organization is less hierarchical, and each product division is responsible for all steps in the product delivery process from R&D to production, sales, and service (Walker, 1993, p. 3; Murray, 1995, p. 3). In 1997, Xerox had more than $18 billion in revenues (see Table F.1), more than 50 percent of which were generated in markets outside the United States. Its facilities (administration, customer service, manufacturing, etc.) are all over the world. Major manufacturing facilities are in Brazil, Canada, China, Egypt, India, Indonesia, Japan, Mexico, the Netherlands, the Philippines, South Korea, Spain, and the United Kingdom. Major U.S. manufacturing locations are El Segundo, California; Oklahoma City, Oklahoma, and Webster, New York.

ENVIRONMENTAL MANAGEMENT HAS EVOLVED FROM COMPLIANCE TO RESOURCE EFFICIENCY AND WASTE ELIMINATION

Environmental management at Xerox is coordinated by the Environment, Health, and Safety (EH&S) group. Its activities have evolved over time—from a

Table F.1

Xerox Total Revenues, Personnel, and R&D Investments

	1997	1996	1995	1994	1993
Total revenues (millions of dollars)	18,166	17,378	16,588	15,084	14,229
Sales	9,892	9,285	8,750	7,823	7,211
Service and rentals	7,268	7,078	6,830	6,255	5,954
Finance income	1,006	1,015	1,008	1,006	1,064
Employees		86,700			
R&D investments (millions of dollars)	1,079	1,044	949	895	883
R&D as a percentage of revenues	5.9	6.0	5.7	5.9	6.2

SOURCES: Xerox, 1996, 1997a, and 1998.

traditional compliance focus to concern about resource use and waste management in addition to compliance. This change came in part because of European customers' concerns over solid-waste generation, the 1990 Pollution Prevention Act, high landfill costs, other financial benefits to the corporation, and Xerox's history of employing the principles of Total Quality Management (TQM) to its business processes (Murray, 1995, pp. 4–5; PCSD, 1997, p. 14). These principles were introduced in 1983. In 1989 Xerox won the prestigious Baldrige Award for quality management.[1]

At its core, TQM emphasizes three elements (Levine and Luck, 1994; Womack and Jones, 1996):

- Identify the organization's customers and what the customers want, now and in the future.

- Identify the processes in the organization that serve the customers and eliminate as much waste—activity that does not add value to the customers—as possible from these processes.

- Monitor performance against the first two goals and continually improve that performance over time.

"Quality" refers to anything the customer values. TQM seeks to eliminate any activity that does not contribute to quality, so defined.

---

[1]TQM is a family of management techniques initially developed in the United States in the 1940s, then refined and expanded in Japan during the 1950s, 1960s, and 1970s, and rediscovered in the United States n the 1980s. The Baldrige Award program was established by Congress in 1987 and named for the late Commerce Secretary Malcolm Baldrige to recognize individual U.S. companies for their achievement and to promote awareness of TQM practices by providing information on successful strategies.

In 1990 the EH&S group developed a portfolio of environmental initiatives called the Environmental Leadership Program (ELP). The primary objective of the ELP was, and is, to improve Xerox's productivity and global competitiveness while minimizing environmental burdens (Xerox, 1997b, p. 3). The five original initiatives of the Environmental Leadership Program were an employee involvement program, a supplies project, a waste-reduction and recycling initiative, a copy cartridge recycling program, and an asset management initiative. These initiatives directly address the second tenet of TQM described earlier—to identify processes that serve the customer and eliminate as much "waste" as possible.[2]

It has led to numerous interrelated activities geared to improve the corporation's resource efficiency and has expanded its initial focus from the design of new products for reuse, remanufacture, and recyclability to all aspects of product delivery (packaging, marketing, services and products procured by Xerox, workplace environments) (Xerox, 1997b, p. 3). A major component of the ELP is the asset recycle management initiative including design-for-environment activities (described in more detail below). In addition to this initiative, in 1993 Xerox established its own corporate goal, supported by CEO Paul Allaire, of "producing waste-free products from waste-free factories"—understanding that the two are interrelated. The definition of "waste-free" for products and factories has been developed by the EH&S group.

These objectives will influence the kinds of investments Xerox makes in environmental technologies, and the kinds of technologies it seeks from outside sources, to address its goals.

R&D IS PRODUCT-ORIENTED

For the last three years investments in R&D have hovered around 6 percent of revenues, higher than the manufacturing average of 3.6 percent. Most of the corporate R&D is focused on product-oriented rather than manufacturing-oriented research. It is prioritized and driven by the strategic business units. A major thrust in the management of Xerox's R&D is the reduction of new products' and services' time-to-market to gain competitive advantage.

Time-to-market is especially important for technology-based businesses because of the uncertainty surrounding future markets and technologies. Companies that can properly identify, and then rapidly respond to, emerging markets will gain competitive advantage. Mark Myers, senior vice president for

---

[2]"Customers" for EH&S in this case are external (the local community, consumers of Xerox products and services) and internal (product line design teams, manufacturing, corporate research and technology, etc.) to the corporation.

Corporate Research and Technology at Xerox, has suggested that since technology innovation is nonlinear, leading companies must be able to identify and respond rapidly to emerging markets, gain early market experience, and then incorporate this experience into future innovations—called radical incrementalism. Having these capabilities is essential to remaining a leader. However, they must be organized—both for existing product systems and for entirely new or radical approaches—to do this if they are going to sustain long-term growth. Because innovating involves more than R&D or technology, R&D managers should forge tight links with many other corporate organizations. A shared strategic vision based on expectations regarding technological and sociological changes can help strengthen these links and can guide R&D investments. As market opportunities are identified, the entire corporation must be able to respond quickly.[3]

To enhance its ability to respond quickly to emerging opportunities, Xerox created a Strategy Council, whose members represent corporate business strategy, the business groups, and corporate research and technology.

This council, which meets quarterly, formulates a strategic vision of emerging customer needs and methods for Xerox to address them. This marks the beginning of the time-to-market process. Business and technology priorities are developed based on these customer needs and emerging market, business, and technology trends. The technology requirements identified in this process are incorporated into the Corporate Research and Technology planning process.

The emphasis on time-to-market means that environmental issues must also be vetted in a time-sensitive manner. As the product realization process accelerates, manufacturing changes must be able to respond quickly as well. Early manufacturing facility planning and pollution prevention through material selection and process design are two tactics for reducing the complexity and time required by the environmental permitting process. However, with rapid product innovation, environmental planning for permitting is quickly becoming a constraint on process cycle times (in some cases, it is already the constraint). Another area affected is safety. The safety approval process for new products, including toxicology testing of new materials, has historically taken on the order of 12 months. Now that some Xerox products turn over in six to 12 months, the safety approval process has been restructured to reduce its time without increasing risk.

R&D is performed at one of the four corporate R&D centers within the Corporate Research and Technology Organization. Each of these centers—Xerox

---

[3]For a more thorough discussion of these ideas, see Myers and Rosenbloom (1996).

PARC, Xerox Research Center Canada, Wilson Research Center, and Rank Xerox—concentrates on different technology areas. Xerox PARC emphasizes longer-term (beyond five years) product-oriented research, most of which is systems and software kinds of work. This research is not *driven* by environmental concerns, although it has an environmental component. Product noise suppression is the primary environmental issue addressed at this center. Xerox Research Center Canada and the Wilson Research Center work closely with the EH&S staff. Xerox Research Center Canada has the lead for new document-processing materials. These include transparencies, inks, toners, specialty chemicals, developers, and photoreceptors. This is where the product-related environmental issues, such as new materials, design-for-environment, and echo toner, are researched. The release of toner from paper is a major issue in the paper industry, which is investing significant resources into recycling. Xerox Research Center Canada also has the lead for studying office operations and recycling issues. Current efforts focus on captive or controlled sites for business services. The Wilson Research Center focuses on shorter-term (three to five years), indirect product support for mainstream development operations. Examples of environmental work at this center include looking at minimizing or eliminating volatile organic compound (VOC) emissions from machines, new manufacturing technologies with lower air or water emissions, and toner particulate size for high-quality printing. Most of the research at the fourth corporate laboratory, Rank Xerox, is oriented toward software development and, as such, has no environmental component. Fuji Xerox research laboratory is an independent laboratory, and the work performed there generally complements the work at the corporate research labs. Overall, Xerox's research and technology development is both product- and process-oriented—with a majority of development going to product-oriented work.

Environmental R&D Is Integral to Overall R&D

EH&S issues are integral to R&D performed by Xerox (a large portion of which is product-oriented). Concern about safety and regulatory compliance has been part of its corporate culture dating to the company's inception. New materials undergo extensive health and safety analyses. And environmentally preferable practices are becoming a source of competitive advantage. Therefore, these issues impact most of Xerox's R&D. A relatively small part of R&D investments are driven by environmental goals exclusive of global safety, regulatory, and environmental product standards.

A combination of regulations, voluntary programs, customer interests, and competitor actions influence the focus of environmental research and product design standards at Xerox. Because Xerox operates facilities and sells equipment worldwide, it is influenced by many different national regulations, product standards, and customer concerns. Because Xerox products last for many years, a long-term forecast (about ten years) of these factors is necessary to establish product requirements. So that these trends can be incorporated into the product realization process as early as possible, the EH&S group has established a Regulatory Tracking Network to provide information on worldwide regulatory trends to the Strategy Council, business groups, R&D groups, and product development teams. This increases the awareness of customer interests to these groups and helps prevent possible delays in the time-to-market of new products (Xerox, 1997b, p. 6).

These regulatory and market-based factors are used to determine product goals or requirements, which are then prioritized into four management categories. These are goals that (1) the company must accomplish for legal reasons, (2) are necessary to meet marketplace requirements, (3) reduce or limit Xerox liability, or (4) contribute to competitive advantage. Depending on the specific issue, designs may exceed regulatory requirements or standards because of trends in the market and long product lifetimes. For example, some toxics are minimized at detection thresholds rather than the higher regulatory thresholds. These priorities are also used to determine environmental technology needs and to make investments in research.

As input into the corporate R&D planning process, a product-oriented technology plan and research needs report was developed by the EH&S group in the 1995 time frame (it was called der Gedanken team). This team reviewed worldwide, customer-driven requirements for Xerox products and identified four environmental technology areas for in-house research. These areas are the following:

- *Energy efficiency* over the product life cycle. Examples include lower energy fusers, improved toner materials, energy-efficient fusing, and more efficient power supplies.

- *Chemical and physical emissions* include technologies to reduce noise in run and standby modes, electromagnetic emissions, and ozone emissions (below the detection threshold).

- *Natural resource conservation.* Included are technologies to enable easy and cost-effective duplex mode, reduce the weight of products for the same productivity, increase the recycled content of plastic parts, reduce the use of nonrenewables, enable the use of recycled paper at no compromise to image quality or duplication rate and reliability, and develop analytic approaches to materials separation.

- *Waste management* technologies and techniques for all life cycle phases. Examples here include low-mass toner, technologies to improve toner efficiency, low-emission disassembly techniques, materials separation technologies, benign degreasing agents and solvents, and solvent avoidance technologies, such as water-based coatings for transparencies, and alternatives to nonrecyclable materials and hazardous materials. Also, comprehensive design techniques for extended life and improved reliability/lower failure rates.[4]

Xerox technologists have established specific targets or goals for each of the areas, which are receiving approximately the same level of investment totaling 1 percent to 2 percent of the overall corporate budget.[5] Most of this research is oriented toward product issues, as opposed to manufacturing-process issues (although there are large investments in emissions-reductions research and technology). Technologies for asset recycle management have been considered in other R&D planning forums and have received lots of attention. Remediation technology issues have been solved. (Xerox developed its own patented remediation technology called Xerox 2-PHASE Extraction™, which it licenses to others. At this point, all that remains in the remediation area is project execution and remediation management streamlining.) Environmental management and research focuses on prevention technologies rather than remediation or control and treatment technologies.

Since the der Gedanken team's experience with environmental technology planning has shown that environmental issues will not receive priority for investments unless they are integrated with other business issues, the EH&S group is engaged in outreach to the research centers, product teams, and manufacturing units to increase awareness of environmental needs.[6] They are marketing the plan to the product units, design teams, and laboratories; publicizing and distributing consumer and regulatory trends data to increase the awareness of these forces; and, where possible, they are linking these technology gaps to opportunities to increase competitive advantage, satisfy customers, etc. The need to comply with regulations provides the most straightforward justification for environmental R&D investments. For more of the proactive environmental needs, the benefits must be related to the bottom line for the product or business units to give such issues higher priority in the

---

[4]Fussler with James (1996, p. 242) notes that Rank Xerox is focusing on ways to deal with nonusable wastes such as foams, seals, painted plastics, toners, and PVC power cords.

[5]These estimates of overall investment are very rough and were provided by Jack Azar, director, EH&S, in a personal interview (1997). The distribution of these investments was estimated by Mark Myers, senior vice president, Corporate Research and Technology (1997).

[6]In 1994, environmental issues were not on the agenda of the R&D planning board. In 1995, the EH&S staff (der Gedanken team) had prepared an environmental technology needs document for R&D planning purposes. These needs were not, however, given high priority by the business units, and as a result very few research areas were funded.

competition for R&D funds. The most successful way to elevate the priority given to an environmental project proposal is to link the attributes to customer interests or the ability to reduce the time-to-market. For example, noise emissions and energy efficiency are environmental issues that are valued by the customer. However, these "win-win" opportunities can be difficult to find. Ultimately, Xerox EH&S personnel hope to integrate the treatment of environmental attributes completely with the other attributes considered in technology planning to increase the level of funding devoted to these technology areas. After approximately five years of working to integrate technology planning, EH&S personnel estimate they are one to two years from full integration of these issues.

Many of Xerox's environmental management activities are process-oriented. Specifically, asset recycle management involves design, manufacturing, and distribution issues. Waste-free factories are almost all process issues. These areas have potentially tremendous payoff to the environment and to the company's bottom line. This strong linkage has ensured that needed investments in this area are made. While product-oriented research represents a large majority of Xerox's R&D investments, the link between environmental features and customer interests and therefore the company's bottom line is not so apparent to those outside the EH&S area. Efforts are under way to increase the awareness at such key organizations as product design units, and the research laboratories.

PROCESS IMPROVEMENTS HAVE LED TO RESOURCE AND COST SAVINGS AND ENVIRONMENTAL BENEFITS

> Waste-free factories are an integral part of the industrial ecology where used parts and material become the feedstocks of future production.
>
> —*Allan Dugan, senior vice president, Corporate Strategic Services (Xerox, undated, p. 7.)*

Asset Recycle Management

Xerox's asset recycle management program is an example of a "win-win" situation. It is where numerous environmental and financial benefits coincide. The asset recycle management program, a major emphasis of Corporate Strategic Services, seeks to maximize the revenue generated by an asset or piece of equipment over its useful life (cradle to cradle). One central corollary of this is to maximize the amount of time an asset, or a piece of equipment, generates revenue. As a result, environmental benefits accrue in addition to the financial benefits, because each component is generating more functionality over its useful life. Consequently, reductions in the generation of solid waste have been tremendous. Upstream savings in energy, material, and air and water emissions should be quite large as well because use of virgin material is much lower. Operationally, asset recycle management has influenced all aspects of the product delivery process from design to production to distribution.

The genesis for this program lay in an interest in velocity management by Al Dugan, senior vice president for Corporate Strategic Services. In 1991, a meeting was called to discuss how management of a very large inventory of machines of high value might be improved.[7] Initially, management options were limited by the lack of information necessary to make decisions about equipment fate. Because the design teams had been disbanded, data on equipment performance or failure modes were not available. Nevertheless, there were potentially large savings given the amount of money that was tied up in the inventory. After the first 12 months of the program, Xerox saved $50 million from changes in logistics practices, inventory policies, and raw materials purchasing alone (Murray, 1995, p. 9). As the asset management options have expanded because of changes in design and delivery practices, the corporation has accrued even more savings. In 1997, Xerox saved *several hundred million dollars* through cost avoidance from material, component fabrication, logistics, and transportation costs alone (Xerox, 1997b, p. 2).

At first, Xerox began asset recycle management with a small project that had high customer interest. It determined what it took in terms of design changes and product delivery process changes to make money from returned copy cartridges. Gradually, asset management concepts were integrated into all projects. Now asset management is changing the relationship between the original equipment manufacturer, supplier, and customer. And it requires new decision criteria and tools, distribution procedures, remanufacturing processes, etc. Xerox has found that extending component life and increasing component commonality are key design tenets. Data on sales volumes, return rates, cost of return, remanufacturing, intermodel conversion potential and design constraints, return on equity for scrap, and other disposal costs are used to make decisions on new designs (Murray, 1995, p. 11). Distribution channels are important for collecting data on failure modes and for collecting returned equipment.

Hugh Smith, Rank Xerox's manufacturing and supply chain environmental manager, put it this way:

---

[7]Fussler with James (1996, p. 240) notes that about this time the original Xerox patents expired, which meant that Xerox was under additional pressure to reduce costs to compete with new entrants, particularly the Japanese companies. He also notes that solid-waste disposal to landfill was becoming more controversial and costly.

> Remanufacturing involves us in supplying solutions and a service. It is a slow process convincing people but our long-term vision, for business and environmental reasons, is to establish the concept of remanufactured products being top quality. If sustainable development is rigorously applied to business, there is no alternative. We are in a pioneering stage, trying to lead demand by educating customers. (Fussler with James, 1996, p. 246.)

Customer relationships are especially important for two reasons— to maximize equipment returns and for product acceptance (equipment as well as consumables). For example, in the case of copy cartridges even though there was high customer support for component reuse, Xerox had to devise incentives to customers to change behavioral patterns and to increase component returns. Eventually, return envelopes with prepaid postage were determined to be the most effective method for copy cartridges. Since the program was initiated in 1991, return rates have risen to almost 65 percent worldwide in 1996 (Xerox, 1997b, p. 19). This practice has been extended to toner bottles, which are more difficult to cost-effectively recycle because of low component value and residual toner that has to be dealt with. After two years of the toner bottle recycle program, return rates have achieved 35 percent (80 percent is the estimated maximum possible return rate) (Xerox, 1997b, p. 19).

Another customer-related issue is the acceptance of remanufactured equipment. Many federal and state procurement regulations specify product characteristics requiring "new" equipment—versus performance characteristics. Even though remanufactured equipment may perform equally (as demonstrated by reliability data), it may be eliminated from procurement because it is not technically "new." International trade rules also have restrictions regarding new and "used" equipment, material content, etc. Individual customers are also reluctant to buy "used" equipment regardless of warranties and customer service guarantees.

Suppliers are included in the asset recycle management process as well. As new products are designed for multiple life cycles through remanufacture, it is important to bring in the expertise (and the capacity) of equipment suppliers to address these issues. Therefore, Xerox engineers work closely with the suppliers in the design process to improve product designs for remanufacture, to encourage the use of recycled material, and to improve labeling and to develop better materials recognition systems for sorting. In fact, as part of its contractual relationship with its suppliers Xerox requires compliance with applicable EH&S regulations, products that do not contain or are not manufactured with ODS, packaging that is free of toxic heavy metals, plastics marking, and cooperation to achieve the environmental leadership driven goals for product design (such as asset recycle management). In this way Xerox communicates market environmental requirements (for example, prohibited materials such as polybrominates) to its suppliers (Xerox, 1997c).

Most of the asset recycle management accomplishments have been done through organizational and process changes. To a lesser extent internal technology development and outside technology consultation has also been employed. Technology has not been a real constraint. However, if Xerox is to meet its goal of 100 percent reuse and recycle then certain technology issues must be addressed. The technology and management issues of asset management fall into five categories: (1) dealing with the return of equipment and supplies, (2) sorting the equipment and material, (3) establishing decision processes on equipment fate, such as reuse or recycle, (4) increasing the use of recycled materials, and (5) designing products for multiple life cycles. The specific technology needs include advanced sensor technology to monitor performance and failure modes; disassembly techniques, including more automated ones; decision analysis tools for component fate; recyclable durable plastics; and sensors and control systems for energy efficiency. Outside of a handful of universities that are looking into disassembly issues (Rochester Institute of Technology, Stanford, Carnegie Mellon, Rhode Island), very little work is being done in these areas.

Waste-Free Factory

The technology issues surrounding Xerox's Waste-Free Factory goals are similar to its asset recycle management initiative. Reflecting its TQM culture, all waste, hazardous and nonhazardous, is treated as a sign of an inefficient product, material, or process that affects Xerox's stakeholders—customers, employees, and residents of nearby communities. The Waste-Free Factory goals were initiated in 1993. A definition of Waste-Free in manufacturing was crafted by EH&S personnel for use in self-assessments by the factory. When specific targets are achieved in nine areas—energy conservation, strategic planning, environmental communications, environmental leadership, use of postconsumer materials, air emissions, solid-waste generation, hazardous waste emissions, and water emissions—a factory is determined to be waste-free (Xerox, 1997b, p. 12). For example, if a factory is operating within 90 percent of the estimated theoretical optimum for energy use, it is considered waste-free in that area. Factories that decrease water use by 50 percent, decrease air emissions by 90 percent, increase utilization of postconsumer materials to 25 percent of material purchases, and decrease municipal, hazardous, and chemical waste by 90 percent over their selected baseline are considered waste-free. Experience to date has shown that the self-assessments are useful tools for identifying competitive cost saving investment opportunities that improve resource utilization or prevent pollution as well.

Most of the improvements to date have been achieved either through house-keeping changes or by looking for opportunities to change practices. For example, solid waste was reduced simply by looking for a market for used materials such as platen glass. When new technological approaches are required to achieve these goals, the decision on whether or not to develop them internally is made using the same criteria as for any other technology—if it is unique to Xerox or is critical to its competitive position, the technology will be developed internally. Otherwise outside sources are sought. In one instance, Xerox PARC teamed with a PARC spin-off, dpiX, to identify a technology solution to the treatment of electronics manufacturing wastewater containing hydrofluoric acid and heavy metals. The team found a suitable process (calcium chloride–based) that would allow treatment of the wastewater on-site rather than transporting it off-site for treatment and disposal. The calcium chloride–based process precipitates out the hazardous materials and encapsulates them in a filter cake of nontoxic calcium fluoride (Xerox, 1997b, p. 7). Examples of technology needs for waste-free factories that are researched internally have to do with toner manufacturing and solvent use in photo-receptor manufacturing. The risks associated with new technological approaches are usually managed by introducing incremental changes, for example testing at one facility before introducing to all. There is also a strong preference for employing proven technologies to avoid the risk associated with the first application.

Other factors that can influence the success of waste-free goals are recycling-market fluctuations and plastic components logistical challenges.

Material cost savings are not the only objective of Xerox waste-free factories. In terms of environmental permits, the manufacturing line (for high-end products) has been relatively stable so permits didn't constrain process changes. Product lines today, however, are experiencing such rapid technological change that the permitting process could potentially constrain Xerox's ability to make necessary manufacturing process changes. Certain product lines are nearly at this point. Waste-free factory goals may help reduce permit complexity and this combined with upfront environmental planning may help reduce the time it takes to make manufacturing process changes.

Technology needs and opportunities for environmental purposes have evolved at Xerox. Ten years ago, its primary focus was on compliance issues. However, resource efficiency and waste minimization have become more visible throughout the corporation. Improvement in resource efficiency and waste generation have been, and can be, made without significant investments in new technology. For the remaining technology issues, Xerox generally invests small amounts on product-oriented environmental features and a smaller amount on process work. It also relies on outside sources for more generic process-oriented needs.

Because of the complexity and uncertainty surrounding the environmental impacts of products, and because product environmental attributes are neither universally defined nor well-defined, consumer communication is complicated. In some cases it may even draw attention to issues previously ignored, or open the company to competitor attacks. As a result, environmental product attributes are to some extent a two-edged sword. These are some of the issues that must be addressed in moving away from product-oriented systems to functionally oriented systems.

FEDERAL ACTIONS THAT COULD HELP XEROX ADDRESS ITS ENVIRONMENTAL ISSUES

The following are federal policies that Xerox personnel suggested could accelerate or expand its investments in environmental technologies.

- Provide strategic direction on environmental issues, such as the leadership that occurred with ODS in the past. This will improve industrial long-range environmental planning and will reduce the risk and uncertainty of investments made in such areas.

- Promote basic *and applied* research in the sciences at universities to ensure the continuation of knowledge generation as well as to ensure that a skilled workforce will be available to actuate the coming wave of technological advances. Substantially increase, by a factor of two, federal funding of science and technology. Xerox and other companies are emphasizing information technology and biotechnology and are less interested in the physical sciences.

- Support education and information dissemination for consumers so that they can make wise choices with respect to environmental concerns. Increase consumers' awareness of environmental benefits to such product features as duplex mode, recycled content, and remanufactured content. Help create consistent definitions of environmentally preferable products.

- The Energy Star labeling initiative is a good program and could be improved with greater consumer education. Incorporate consumer preferences into future environmental product standards in general.

- Public-private partnerships in the technology areas of energy efficiency technologies (e.g., standby mode and rapid on/off, which is a sensors, control technology, and materials issue), recyclable plastics for durable goods, recyclable elastomers, remanufacturing and disassembly techniques (e.g., degreasing agents and techniques, materials sorting technology, component characterization), product design analysis tools (e.g., for life-extension or recycle decisions), product life-prediction analyses, and the development of life-cycle assessment data bases. Cooperate on recycling infrastructure development. Many of these technology areas have broad applications. Currently Xerox has some university partnerships in a few of these areas but none with the federal government. Xerox personnel suggested that PNGV and the Canadian pulp and paper initiative were two examples of good public-private collaborative efforts. While public-private cooperation was welcomed, a fundamental difficulty identified was that the federal government has a harder time changing research priorities than private industry does.

- Create markets and incentives for continuous environmental improvement. Use federal, state, and local procurement agencies to stimulate markets for environmentally preferable products and add flexibility to procurement regulations so that remanufactured products can be procured. Help develop the necessary infrastructure for recycled materials.

- Ensure that patent laws are enforced and strengthened as more companies move to the globally distributed model. Patents will become a form of currency.

- Harmonize U.S. environmental and trade policies with the world market. Create environmental policies that lead the world, but not by too much. Work to remove international trade restrictions on remanufactured equipment. Eliminate barriers to seeking new technologies internationally.

- Fund total systems modeling of the economics and resource flows of product systems. Information from a total systems model could be employed in the design of new products to provide guidelines for design trades. For example, the decision to design a recyclable product versus a longer-lived product may include factors beyond the scope of an individual company.

SUMMARY AND CONCLUSIONS

Markets, regulations, product standards, and competitor actions are the major drivers of environmental investments. If customers demand environmentally beneficial features, or if offering these products improves a company's competitive position, these features will be provided. Xerox's own initiatives and programs that seek greater resource efficiency and effectiveness create potential demand for new environmental technologies. Specifically, the asset recycle management program and the waste-free products produced in waste-free factories goals influence its demand for technology, although many of the improvements to date have been achieved without implementing large amounts of new technology.

Environmental compliance, health, and safety issues are integral to all R&D performed by Xerox (a large portion of which is product-oriented). These issues impact most of Xerox's R&D, which is directed toward features desired by customers. A relatively small part of its R&D investments are exclusively driven by environmental goals outside of global safety, regulatory compliance, and product standards. The majority of in-house research is product- and not process-focused. University-based research, which is limited in the area of disassembly and remanufacturing, is also tapped.

Xerox personnel have identified four categories of technology needs for research investments. These are energy efficiency technologies, natural resource conservation technologies (two-sided copying, use of recycled materials, etc.), chemical and noise emission reduction technologies, and life-cycle waste management technologies. Investments have been made in these categories in roughly equal proportions. In terms of its emphasis, Xerox is primarily interested in avoidance technologies. It has no technology need for additional remediation technologies.

Xerox personnel have several suggestions for federal policy actions to improve investments in environmental technologies. These recommendations include social or behavioral, regulatory, and investment policies. Policies that maintain the scientific infrastructure, educate consumers, train the workforce, and provide frameworks that bring all stakeholders or constituencies together to solve societal problems are desired. In order for the United States to be innovative in general, it must have the capability to address problems collectively and commercialize the requisite products and services. Many personnel suggested policies to create favorable conditions for environmental products in the long term. Partnerships are one of many possible approaches to increasing the investments made into environmental technologies. These interest areas desired by Xerox do not fit neatly into the mission of any one federal agency.

This suggests that such policies as product standards, consumer education and labeling, and affirmative procurement, which influence demand and consumer preferences will help create a demand-pull for environmental technology. Regulations are another forcing function for environmental technology investments. They are useful because they identify clear and direct priorities to industry. However, they can be costly if they are not targeted so that environmental outcomes are improved at reasonable expense. Other factors, such as the price of materials, price of energy, the availability of recyclable materials, and the cost of waste disposal will also encourage resource efficiency and the necessary technology investments.

Azar, Jack, Director, Environment, Health, and Safety, Xerox, May 1997 and May 1998.

Beaver, Earl, "The Monsanto $1 Million Challenge," presentation to the World Congress of Chemical Engineering, July 1996.

_____, Director, Waste Minimization, Monsanto, interviews, December 1997 and April 1998.

Bower, Joseph, and Clayton Christensen, "Disruptive Technologies: Catching the Wave," *Harvard Business Review*, January-February 1995, pp. 43–53.

Brandenburger, Adam, and Barry J. Nalebuff, "Inside Intel," *Harvard Business Review*, November-December 1996, pp. 168–175.

Brodsky, Phil, Vice President, Corporate Research and Environmental Technology, Monsanto, interview, March 1997.

Brodsky, Phil, and Earl Beaver, Monsanto, interview, May 1997.

Byrne, John A., "Strategic Planning," *Business Week*, August 26, 1996, pp. 46–52.

Carberry, John, Director of Environmental Technology, DuPont, interviews, February and April 1997.

Chapman, Rob, former PNGV chairman, RAND presentation, May 1997.

Chatterji, Deb, "R&D Management and the Environmental Imperative," mimeo; British Oxygen, 1993.

Council on Competitiveness, *Endless Frontier, Limited Resources: US R&D Policy for Competitiveness*, Washington, D.C., April 1996.

Dertouzos, Michael L., Robert M. Solow, and Richard Keith Lester, *Made in America: Regaining the Productive Edge*, Cambridge, Mass.: MIT Press, 1989.

Dodgson, Mark, and Roy Rothwell, eds., *The Handbook of Industrial Innovation*, Brookfield, Vt.: Edward Elgar Publishing Company, 1994.

Dumond, John, Rick Eden, Douglas McIver, and Hy Shulman, *Maturing Weapon Systems for Improved Availability at Lower Costs*, Santa Monica, Calif.: RAND, MR-338-A, 1994.

DuPont, "DuPont Announces Plans To Divest Its Conoco Energy Operations," news release, May 11, 1998b.

_____, "DuPont President and CEO Addresses Yale School of Management," news release, March 6, 1996c.

_____, "Profitable Growth Will Drive Success," news release, January 26, 1996b.,

_____, "Randy Guschl Was Keynote Speaker at the 50th Anniversary Symposium at Penn State's Applied Research Laboratory," news release, June 15, 1995a.

_____, *1996 Annual Report to Shareholders*, 1997a.

_____, *1997 Annual Report to Shareholders*, 1998a.

_____, news release, September 20, 1995b.

_____, *DuPont Safety, Health, and the Environment: 1995 Progress Report*, 1996b.

_____, *DuPont Safety, Health and the Environment: 1996 Progress Report*, 1997b.

Economist, The, "Genetic Engineering: The Year of the Triffids," April 26, 1997, pp. 80–82.

Energy Information Agency, *Annual Energy Review Historical Data: Consumption by End-Use*, accessed on www.eia.doe.gov November 1998.

Environmental Law Institute, *Research and Development Practices in the Environmental Technology Industry*, Environmental Law Institute Project #941735, 1997.

Environmental Protection Agency, *Report and Recommendations of the Technology Innovation and Economics Committee: Permitting and Compliance Policy: Barriers to US Environmental Technology Innovation*, EPA 101-N-91-001, National Advisory Council for Environmental Policy and Technology, Washington, D.C., January 1991.

_____, *Improving Technology Diffusion for Environmental Protection: Report and Recommendations of the Technology Innovation and Economics Committee*, EPA 130-R-92-001, October 1992.

_____, *Transforming Environmental Permitting and Compliance Policies to Promote Pollution Prevention: Removing Barriers and Providing Incentives to Foster Technology, Innovation, Economic Productivity, and Environmental*

Protection, EPA 100-R-93-004, National Advisory Council for Environmental Policy and Technology, Washington, D.C., April 1993.

_____, *TRI Summary Report*, "Table 1-21—Top 10 Parent Companies with the Largest Total Releases, 1994." Available at http://www.epa.gov, accessed September 1996.

_____, "Profile of the Electronics and Computer Industry," EPA Office of Compliance Sector Notebook Project, Office of Enforcement and Compliance Assurance, EPA/310-R-95-002, September 1995. Available at http://www.epa.gov.

_____, *Intel Final Project Agreement Response to Comments*, 1996. Available at http://www.epa.gov.

_____, *Remediation Technologies Development Forum Fact Sheet*, EPA 542-F-97-012, November 1997.

Feder, Gershon, and Dina Umali, "The Adoption of Agricultural Innovations: A Review," *Technological Forecasting and Social Change*, Vol. 43, Nos. 3/4, May/June 1993.

Fiksel, Joseph, ed., *Design for Environment: Creating Eco-Efficient Products and Processes*, New York: McGraw-Hill, 1996.

Florida, Richard, "Technological Policy for a Global Economy," *Issues in Science and Technology*, Spring 1995, pp. 49–56.

_____, "Lean and Green: The Move to Environmentally Conscious Manufacturing," *California Management Review*, Vol. 30, No. 1, Fall 1996, pp. 98–101.

France, Wayne, Manager of Environmental, Health, and Safety, General Motors, personal communication, March 1997.

Fritsch, Peter, "Monsanto May Shed Its Chemical Unit," *Wall Street Journal*, October 11, 1996, pp. A3–A4.

Fusfeld, Herbert, *Industry's Future*, American Chemical Society, Washington, D.C., 1994.

Fussler, Claude, with Peter James, *Driving Eco-Innovation*, Washington, D.C.: Pitman Publishing, 1996.

Gibbs, W. Wayt, "The Price of Silence: Does Profit-Minded Secrecy Retard Scientific Progress?" *Scientific American*, November 1996, pp. 15–16.

Grant, Linda, "Monsanto's Bet: There's Gold in Going Green," *Fortune*, April 14, 1997, pp. 116–118.

Green, Kenneth, Andrew McMeekin, and Alan Irwin, "Technological Trajectories and R&D for Environmental Innovation in UK Firms," *Futures*, Vol. 26, No. 10, 1994.

Holliday, Chad, President and CEO of DuPont, remarks at the Biotechnology Panel Discussion of the World Economic Forum, Davos, Switzerland, February 3, 1998.

Hamilton, Martha M., "Linking Business's Climate and Earth's," *Washington Post*, January 12, 1997, p. H1.

Hamilton Martha M., and Justin Gillis, "$33 Billion Deal to Form Life-Sciences Giant," *Washington Post*, June 2, 1998, p. C1.

Harpole, Tom, "The Cold Business Logic of Sustainability," *Monsanto Magazine*, No. 1, 1997, pp. 20–23.

Hart, Stuart, "Beyond Greening: Strategies for a Sustainable World," *Harvard Business Review*, January-February 1997, pp. 66–77.

Hutcheson, G. Dan, and Jerry D. Hutcheson, "Technology and Economics in the Semiconductor Industry," *Scientific American*, January 1996, pp. 54–62.

Institute for the Future, *The Future of America's Research-Intensive Industries*, Menlo Park, Calif.: R-97. August 1995, p. 12.

Intel Corporation, *1995 Annual Report to Shareholders*, 1996a.

_____, *Environmental Health and Safety at Intel*, March 1996b.

_____, "Intel and Rochester Electronics Sign Technology License Agreement," press release, May 13, 1996c.

_____, *1997 Annual Report to Shareholders*, 1998.

International Cooperative for Ozone Layer Protection, *The International Cooperative for Ozone Layer Protection: 1990–1995*, ICOLP brochure, 1996.

Kirkpatrick, David, "Intel's Amazing Profit Machine," *Fortune*, February 17, 1997, pp. 60–72.

Kline, Stephen J., "Innovation Is Not a Linear Process," undated.

Kling, James, "Could Transgenic Supercrops One Day Breed Superweeds?" *Science*, Vol. 274, October 11, 1996, pp. 180–181.

Krol, Jack, "The Chemical Industry—Indispensable in the 21st Century," speech presented to Pittsburgh Chemical Day, Pittsburgh, Pa., April 15, 1997.

Krewinghaus, Bruce, Shell Development Corp., personal communication, February 1997.

Labate, John, "Chipmakers Have a Drinking Problem," *Fortune*, Vol. 132, No. 16, September 18, 1995, p. 32.

Lachman, Beth, Tom Anderson, Robert Lempert, Susan Resetar, *Technology for a Sustainable Future Ideas: A Summary of Stakeholder Workshop Discussions*, Santa Monica, Calif.: RAND, RP-417, 1995.

Lenz, Allen, and John LaFrance, *Meeting the Challenge: US Industry Faces the 21st Century, The US Chemical Industry*, Washington, D.C.: Office of Technology Policy, U.S. Department of Commerce, January 1996.

Levine, Arnold, and Jeff Luck, *The New Management Paradigm: A Review of Principles and Practices*, Santa Monica, Calif.: RAND, MR-458-AF, 1994.

Link, Albert, and Laura Bauer, *Cooperative Research in US Manufacturing: Assessing Policy Initiatives and Corporate Strategies*, Lexington, Mass.: Lexington Books, 1989.

Lissoni, F., and J. S. Metcalfe, "Diffusion of Innovation Ancient and Modern: A Review of the Main Themes," in *The Handbook of Industrial Innovation*, Mark Dodgson and Roy Rothwell, eds., Brookfield, Vt.: Edward Elgar Publishing Company, 1994.

Loutfy, Rafik, Vice President, Strategy and Innovation for Corporate Research and Technology, Xerox, interview, July 1998.

Magretta, Joan, "Growth Through Global Sustainability: An Interview with Monsanto's CEO, Robert B. Shapiro," *Harvard Business Review*, January-February 1997, pp. 79–88.

Mazur, Allan, *A Hazardous Inquiry: The Rashomon Effect at Love Canal*, Cambridge, Mass.: Harvard University Press, 1998.

McGraw-Hill and EPA, "Environmental Champion Awards for 1995 Announcement," 1995.

McManus, Terry, Manager of Corporate Environmental Affairs, Intel, interview, February 5, 1997.

Meadows, Donella, "Our Footprints Are Treading Too Much Earth," *Charleston Gazette*, April 1, 1996, in Hart (1997), p. 68.

Merges, Robert, and Richard R. Nelson, "On the Complex Economics of Patent Scope," *Columbia Law Review*, Vol. 90, No. 4, May 1990, pp. 839–916.

Miller, Joseph A., "Discovery Research Re-Emerges in DuPont," *Research Technology Management*, January-February 1997, pp. 24–26.

Mohin, Timothy, presentation at the National Pollution Prevention Roundtable Conference, Washington, D.C., April 1996.

_____, Government Affairs Manager, Environment, Health, and Safety, Intel, interview, April 4, 1997.

Monsanto, "Monsanto Announces Second $1 Million Challenge for Waste Recovery Solution," press release, November 15, 1995.

_____, news brief, *Monsanto Magazine*, Number 4, 1996, p. 27.

_____, *1996 Monsanto Annual Report to the Shareholders*, 1997a.

_____, *1996 Environmental Annual Review: Pursuing Sustainability*, 1997b.

_____, "The Cold Business Logic of Sustainability," *Monsanto Magazine*, No. 1, 1997c, pp. 20–23.

_____, "Life Sciences Business to Retain Monsanto Name; Solutia Inc. Selected as Name of Chemical Spin-off," press release, August 18, 1997d.

_____, "Monsanto Shareowners Approve Spinoff Of Chemicals Business," press release, August 18, 1997e.

_____, *1997 Monsanto Annual Report to the Shareholders*, 1998a.

_____, *1997 Environmental Annual Review: Pursuing Sustainability*, 1998b.

_____, *1997 Report on Sustainable Development, Including Environmental, Safety, and Health Performance*, March 1998c.

_____, *Sustainable Development Sector* on http://www.monsanto.com.

Murray, Fiona, *Xerox: Design for the Environment*, Harvard Business School Case Study, 9-794-022, rev. May 4, 1995.

Myers, Mark, Senior Vice President, Corporate Research and Technology, Xerox, interview, March 1997.

Myers, Mark, and Richard Rosenbloom, "Rethinking the Role of Research," *Research Technology Management*, May-June 1996, pp. 14–18.

Nanda, Ashish, and Christopher Bartlett, *Intel Corporation—Leveraging Capabilities for Strategic Renewal*, Harvard Business School Case Study, HBS 9-394-141, March 9, 1994.

National Science and Technology Council, *Technology for a Sustainable Future*, Washington, D.C.: Government Printing Office, 1994.

_____, *Bridge to a Sustainable Future*, Washington, D.C.: Government Printing Office, 1995.

National Science Board, *Science and Engineering Indicators 1996*, NSB 96-21, Washington, D.C.: Government Printing Office, 1996.

National Science Foundation, *Research and Development in Industry: 1994*, NSF 96-304, Arlington, Va., 1996.

Office of Technology Assessment, *Industry, Technology, and the Environment: Competitive Challenges and Business Opportunities*, OTA-ITE-586, Washington, D.C.: Government Printing Office, January 1994.

Paugh, Jon, and John LaFrance, *Meeting the Challenge: US Industry Faces the 21st Century: The US Biotechnology Industry*, Washington, D.C.: Office of Technology Policy, U.S. Department of Commerce, September 1997.

Pelley, Janet, "Environmental R&D Shifts to Pollution Prevention," *Environmental Science and Technology News*, Vol. 31, No. 3, 1997, pp. 138A–141A.

Perrings, Charles, "Sustainable Livelihoods and Environmentally Sound Technology," *International Labour Review*, Vol. 133, No. 3, 1994, pp. 307–326.

Piasecki, Bruce, *Corporate Environmental Strategy: The Avalanche of Change Since Bhopal*, New York: John Wiley and Sons, 1995.

Port, Otis, with Andy Reinhardt, Gary McWilliams, and Steven V. Brull, "The Silicon Age? It's Just Dawning," *Business Week*, December 9, 1996, pp. 148–152.

Porter, Michael E., and Claas van der Linde, "Green and Competitive: Ending the Stalemate," *Harvard Business Review*, September-October 1995, pp. 120–134.

President's Committee of Advisors on Science and Technology, "Federal Energy Research and Development for the Challenges of the Twenty-First Century," Report of the Energy Research and Development Panel, Washington, D.C., November 1997.

President's Council on Sustainable Development, *Proceedings of the Workshop on Extended Product Responsibility: October 21–22, 1996*, Washington, D.C., February 1997.

Rea, David R., Vice President, DuPont R&D, "Globalization of R&D: Overview of Trends," presentation to the Council for Chemical Research Annual Meeting, Pittsburgh, Pa., October 8–10, 1995.

Reinhardt, Andy, Ira Sager, and Peter Burrows, "Intel," *Business Week*, December 22, 1997, pp. 70–77.

Reisch, Marc S., "Holliday Blazes His Own Trail," *Chemical and Engineering News*, December 14, 1998, pp. 21–25.

Rejeski, David, "Forum: Getting Into the Swing," *Technology Review*, January 1997, pp. 56–67.

Rice, Faye, "Who Scores Best on the Environment," *Fortune*, July 26, 1993.

Rich, Laurie, "How Environmental Pressures Are Affecting the R&D Function," *Research Technology Management*, September-October 1993, pp. 16–23.

Rogers, Karen Keeler, "Fields of Promise: Monsanto and the Development of Agricultural Biotechnology Part One: Setting the Course," *Monsanto Magazine*, No. 4, 1996, St. Louis, Mo., pp. 4–9.

Rogers, Karen Keeler, "Fields of Promise: Monsanto and the Development of Agricultural Biotechnology Part Two: Creating the Future," *Monsanto Magazine*, No. 1, 1997, St. Louis, Mo., pp. 14–19.

Roome, Nigel, "Business Strategy, R&D Management and Environmental Imperatives," *R&D Management*, Vol. 24, No. 1, 1994.

Rosenberg, Nathan, *Exploring the Black Box: Technology, Economics, and History*, New York: Cambridge University Press, 1994.

Rushton, Brian M., "How Protecting the Environment Impacts R&D in the United States, *Research Technology Management*, May-June 1993, pp. 13–21.

Rushton, Brian M., personal communication, February 1998.

Sanders, Lisa, "Going Green with Less Red Tape," *Business Week*, September 23, 1996, pp. 75–76.

Schmidheiny, Stephan, *Changing Course*, Cambridge, Mass.: MIT Press, 1992.

Schneiderman, Howard, "Managing R&D: A Perspective from the Top," *Sloan Management Review*, Summer 1991, pp. 53–58.

Sciance, Tom, remarks presented to the CMA Manufacturing Committee, July 27, 1994.

_____, personal communication, September 1996.

Sharp, Margaret, "Innovation in the Chemicals Industry," in Mark Dodgson and Roy Rothwell, eds., *The Handbook of Industrial Innovation*, Brookfield, Vt.: Edward Elgar Publishing Company, 1994.

Smart, Bruce, ed., *Beyond Compliance: A New Industry View of the Environment*, Washington, D.C.: World Resources Institute, 1992.

Smith, Douglas, and Robert C. Alexander, *Fumbling the Future*, New York: Morrow, 1988.

Solutia, *1997 Annual Report*, St. Louis Mo., 1998.

Stoneman, Paul, *The Economic Analysis of Technology Policy*, Oxford, U.K.: Clarendon Press, 1987.

Sundbo, John, "Innovation Theory: Three Paradigms in Innovation Theory," *Science and Public Policy*, Vol. 22 No. 6, December 1995, pp. 399–410.

Thurow, Lester, "Needed: A New System of Intellectual Property Rights," *Harvard Business Review*, September-October 1997, pp. 95–103.

U.S. Bureau of the Census, *Pollution Abatement Costs and Expenditures, 1991*, MA200(91)-1, Washington, D.C.: Government Printing Office, 1993.

U.S. Department of Energy, Office of Industrial Technology, Industries of the Future Program, *Chemical Industry Profile*. Accessed on http://www.iot.doe.gov/chemicals/page11.html on November 1998.

Utterback, James, *Mastering the Dynamics of Innovation: How Companies Can Seize Opportunities in the Face of Technological Change*, Boston, Mass.: Harvard Business School Press, 1994.

Van LobenSols, Chris, and David Hawkins, *Comments on the Intel Ocotillo Site XL Project (Project) Draft Final FPA and Draft Air Permit*, Natural Resources Defense Council, 1996.

Walker, Wayne, *Recovering the Fumbles and Organizing for the Future: Xerox Integrates R&D into Corporate Strategy with Pioneering Research and Restructures to Become a Learning Organization—with Lessons for Military Acquisition*, Santa Monica, Calif.: RAND, P-7802, 1993.

Wall Street Journal, "Intel Shifts Focus, Concentrating Now on Original Research," August 26, 1996.

Walley, Noah, and Bradley Whitehead, "It's Not Easy Being Green," *Harvard Business Review*, May-June 1994, pp. 46–52.

White House Press Announcement, Office of the Vice President, *Toxics: Expanding the Public's Right to Know*, April 21, 1998.

White, Robert, "It's Morning in American Industry," *Issues in Science and Technology*, February/March 1997, p. 68.

Womack, James P., and Daniel T. Jones, *Lean Thinking*, New York: Simon and Schuster, 1996.

Woodfin, Max, "Bugging Out," *Sierra*, January-February 1997, p. 27.

World Commission on Environment and Development (WCED), *Our Common Future*, Oxford, U.K.: Oxford University Press, 1987.

Xerox Corporation, *1995 Annual Report to Shareholders*, 1996.

_____, *1996 Annual Report to Shareholders*, 1997a.

_____, *Environment, Health and Safety Progress Report 1996: Final Draft*, May 1997b.

_____, *Xerox Global Purchasing: Supplier Environment, Health, and Safety Requirements*, EH&S 1001, December 7, 1997c.

_____, *1997 Annual Report to Shareholders*, 1998.

_____, *Xerox Products and the Environment: Products for a Sustainable Future*, undated, p. 7.